T0295415

Sea Levels and Coastal Boundaries

Sea Levels and Coastal Boundaries

George M. Cole

Published by John Wiley & Sons, Inc., Hoboken, New Jersey.
Published simultaneously in Canada.

For general information on our other products and services or for technical support, please contact our Customer Care Department within the United States at (800) 762-2974, outside the United States at (317) 572-3993 or fax (317) 572-4002.

Wiley also publishes its books in a variety of electronic formats. Some content that appears in print may not be available in electronic formats. For more information about Wiley products, visit our web site at www.wiley.com.

Library of Congress Cataloging-in-Publication Data
Names: Cole, George M., author.
Title: Sea levels and coastal boundaries / George M. Cole.
Description: First edition. | Hoboken, N.J. : Wiley, 2024. | Includes
   index.
Identifiers: LCCN 2024007503 (print) | LCCN 2024007504 (ebook) | ISBN
   9781394216888 (hardback) | ISBN 9781394216895 (adobe pdf) | ISBN
   9781394216901 (epub)
Subjects: LCSH: Hydrographic surveying. | Topographical surveying. | Water
   boundaries. | Sea level.
Classification: LCC VK591 .C65 2024 (print) | LCC VK591 (ebook) | DDC
   526.9/9—dc23/eng/20240323
LC record available at https://lccn.loc.gov/2024007503
LC ebook record available at https://lccn.loc.gov/2024007504

Cover Design: Wiley
Cover Image: Courtesy of George M. Cole

Set in 9.5/12.5pt STIXTwoText by Straive, Chennai, India.
SKY10078301_062524

Contents

Preface

Humans have always been drawn to the sea and the other great surface waters of the Earth as a source of food, moderate climate, and a medium for transportation and recreation. As evidence of the early attraction of mankind to the sea, remains of some of the earliest human settlements in the United States have been found miles offshore in the Gulf of Mexico. With the limited infrastructure associated with early human life, the dynamic nature of the shoreline, due to long-term sea level change, was not a concern for such settlements. Modern humans are drawn to water for the same reasons as those associated with those early settlements.

Today, it is estimated that over a third of the population and three-fourths of the world's megacities are located in coastal zones. With the more elaborate infrastructure associated with contemporary human development, the constantly moving shorelines associated with long-term sea level change are a topic of considerable interest and concern. As a result, an objective look at long-term sea level change is the primary topic of this writing.

Associated with the issue of long-term sea level change is that of littoral and riparian boundaries and how sea level changes are addressed with such boundaries. The waters of the seas and other major waterbodies have long been considered a public commons. As a result, the boundaries between such waters and the bordering uplands have a special significance as the line between public and private interests. Those lines have been the subject of legal codes since the early Roman civilization. As an example, the early civil code of that era, known as the Institutes of Justinian, reflected the status of the sea as a commons and also defined the boundary between the sea and bordering uplands. This is illustrated by the following translation of a portion of that code.

> The sea shore, that is, the shore as far as the waves go at furthest, was considered to belong to all men . . . . The sea shore extends as far as the greatest winter floods runs up.

Despite the early recognition of the boundary between the sea and bordering uplands, such boundaries remain today as among the most complex in human society due to the complexity associated with the constant movement of the edge of the water.

This writing is an examination of such sea level dynamics and how society deals with the constantly changing sea levels while also protecting the rights of owners of bordering uplands. It begins with a description of how sea level is constantly changing, in both the short and long terms. Then, laws and processes that have evolved for defining the boundary between public waters and bordering uplands are examined for both tidal and nontidal waters. Next, the topic of national and state boundaries in coastal waters and how they are determined is examined. The writing then continues with a review of practices for determining boundaries where shorelines have been altered, followed by an examination of the rights of riparian and littoral property owners and how such rights are defined. The writing concludes with a review of how sea level is changing with interesting conclusions regarding the apparently rapidly changing rates of change along the coastlines of the United States.

Thus, the writing is intended as a guidance in understanding how sea level is changing, how society has defined the boundaries between public waters and bordering uplands as well as national offshore boundaries, and how such boundaries are defined and located. Therefore, the material presented in this writing may be of interest or assistance to practicing boundary surveyors, attorneys, engineers, oceanographers, coastal land managers, landowners, or planners as well as other persons interested in long-term sea level change.

This topic is especially germane at this time due to the public awareness of the possible effects of sea level rise and the frequent sensational coverage of that issue in the popular press. These have increased the need for professionals to understand and be able to deal with such effects. Hopefully, the writing will allow such readers to better understand the overall picture of how sea and other water levels change with time and how littoral and riparian boundaries address such changes. It is also hoped that readers will find the topic as fascinating as it is to the writer even after six decades of practice in this area.

George M. Cole
Chapel Hill, North Carolina
May 2024

1

Sea Level Dynamics

As discussed in the Preface, one of the more complex aspects of water boundaries is the dynamic nature of the land/water interface. A major cause of that variation is constantly changing water level. That fluctuation is due to a variety of causes including the tides, metrological conditions, and global sea level changes. Traditionally, sea level variations are classified by the period of variation, ranging from *surface gravity waves* with periods varying from 1 to 20 seconds; to *seiches* and *tsunamis* with periods of up to an hour; to *astronomic tides* with dominant periods of one-half and one lunar day; to *storm surges* with periods ranging from a few hours to several days; to *long term, apparently nonperiodic trends* caused by geological and climatological effects with periods of thousands of years.

In addition to varying periods, sea level variations also vary considerably in amplitude. Variations range from those associated with seiches and surface waves with amplitudes as small as a few centimeters to tsunamis with amplitudes in the tens of meters.

1.1 Short-Term Sea Level Variation (Other than Tides)

1.1.1 Surface Gravity Waves

Possibly the most noticeable sea level variations are *surface gravity waves* (Figure 1.1), which are generally called either wind waves or swell. *Wind waves* are the effect of wind on water and always travel in the same direction that the wind is blowing. Wind waves continuing for longer than a few hours gain sufficient energy to take on a distinct character known as *swell*, which move across open areas of water even though not under the influence of the wind. Wind waves generally have periods from 1 to 15 seconds. Swell has longer periods,

Sea Levels and Coastal Boundaries, First Edition. George M. Cole.

Figure 1.1 Surface Gravity Waves.

generally between 12 and 25 seconds, and appears less steep than wind waves. Also, swell will not normally break in open water, while wind waves will often break.

The height of waves is usually expressed as the vertical distance from peak to trough. Since there can be considerable differences in height between individual waves in an area, another measurement used to describe heights is the *significant wave height*. That measure is the average height of the highest one-third of waves over an observational period, generally about 20 minutes. Wave heights up to 30 m have been measured.

1.1.2 Seiches

Seiches are the periodic change in sea level that occur in enclosed waters and which are set in motion by some disturbance such as a strong wind, atmospheric pressure changes, or boat traffic. Basically, seiches represent water sloshing back and forth within an enclosed basin with periods ranging from a few minutes to a few hours depending upon the size of the basin. The amplitude of seiches is generally in the tenths of a foot range or less (Figure 1.2).

Knowledge of seiches is important for the design and operation of harbors and other areas where berthing of deep draft vessels occur. In addition, knowledge of

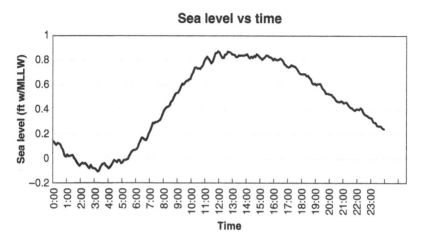

Figure 1.2 Prominent Seiche Superimposed on Tidal Variation for Typical Day, Isla Magueyes, Puerto Rico. *Source:* www.tidesandcurrents.noaa.gov/NOAA/Public Domain.

these variations is important in tidal studies since they can easily distort tidal measurements.

1.1.3 Storm Surges

As previously discussed, surface waves are created by the drag or stress of atmospheric wind on the sea surface. Changes in the level of the sea surface are also related to another atmosphere phenomenon called the *inverse barometer effect*. The combination of the effect of wind drag (which is proportional to the square of the wind speed) and effect of atmospheric pressure (which decreases sea level by one centimeter per millibar) can result in huge sea level surges being generated by storm systems, especially in shallow water bodies.

In simple terms, a mound of water can be produced by a storm moving across a water body. The storm wind moving cyclonically around the storm can push the water causing it to pile up as it approaches the shore. Since tropical storm winds have a counterclockwise motion in the northern hemisphere, the storm surge in that hemisphere is typically greatest in height to the right of an approaching storm. The height of a storm surge in a particular location depends on a number of different factors including storm intensity, forward speed, angle of approach to the coast, central pressure, and the shape and bathymetric characteristics of coastal features such as bays and estuaries, and the width and slope of the continental shelf. A shallow slope will potentially produce a greater storm surge than a steep shelf. As a result, a storm approaching an area such as

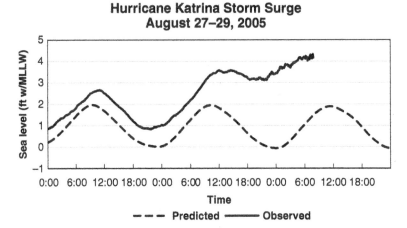

Figure 1.3 Storm Surge of Hurricane Katrina as Observed in Biloxi, MS (Note that the gauge was lost near the height of the storm surge). *Source:* NOAA/https:// tidesandcurrents.noaa.gov/predma2.html, last accessed 13 November 2023.

the Louisiana coastline with a very wide and shallow continental shelf may produce a 20-ft storm surge, while the same hurricane approaching the eastern U.S. coastline with a steeper continental shelf would result in a much smaller surge. Often, storm surges cause greater damage than the winds in a storm (Figure 1.3).

1.1.4 Tsunamis

A *tsunami* is a series of waves created by a displacement of the water column caused by an undersea disturbance of some type. That disturbance may be undersea earthquakes, landslides, volcanic eruptions, or even explosions or meteorite impacts. The period of the waves varies from minutes to greater than an hour, depending on the nature of the disturbance creating the tsunami. Once generated, the waves travel rapidly across the open ocean with a speed equal to the square root of the product of the water depth and the acceleration of gravity ($9.8\,m/s^2$), with speeds often over 620 kmph (IOC 2006). The wavelength may be as long as 80 km. At sea, the wave height may be less than a meter in height, so the waves are virtually unnoticeable at sea. Yet, as the tsunami waves approach shoaling water, the water piles up which creates waves as high as 30 m or more (Figure 1.4).

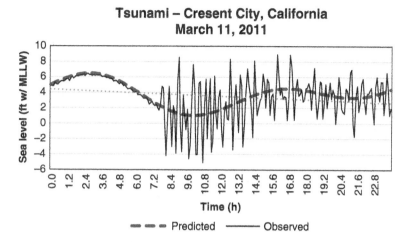

Figure 1.4   Tsunami Waves Resulting from the Japanese Earthquake of 3/10/2011, as Observed in Crescent City, California. *Source:* www.tidesandcurrents.noaa.gov/NOAA/ Public Domain.

1.2   Tidal Variation and Datum Planes

1.2.1   Tidal Cycles

Tides are sea level variations caused by the gravitational forces of the sun and moon as well as solar radiation. Tidal variations are cyclic and follow periodic patterns due to the cyclic nature of those astronomical phenomena. As a result, tides may be distinguished from other types of sea level variations.

The primary driving force of the tides is the gravitational pull of the moon as it rotates around the earth. Due to that force, there is an uplifting of the sea under the moon caused by its gravitational pull on the fluid water. On the side of the earth opposite the moon, the lesser gravitational force due to the greater distance to the moon and the centrifugal force caused by the earth's spin causes a second higher water. Although interrupted by intervening land masses, these two high water waves, with their intervening low waters, follow the moon in its revolution about the earth and represent the primary constituents of the observed tide. Since one-half of the average interval between consecutive transits of the moon is 12.42 hours, the moving high waters generally take the form of a sine wave with a period of that interval.

There is a similar, although somewhat lesser, effect on the level of the seas caused by the gravitational pull of the sun on water on the rotating earth. That wave may be represented as a sine wave with a period of 12.00 hours. Changes in sea level caused by several other relationships between the moon, sun, and earth may also be considered as sine wave constituents of the observed tide. For example, the elliptical orbit of the moon about the earth results in a constituent with a period of 27.55 days with highest water at the time of perigee (when the moon is closest to the earth) and lowest water when the moon is the greatest distance away. Also, there is a constituent period of one year associated with the declination of the sun.

When the constituent cycles associated with the cycles of the moon and sun are "in phase" (when the peaks are occurring at approximately the same time), tides with greater than normal ranges occur. Such is the case twice a month near the time of the new and full moon when the earth, moon, and sun are in a line. At those times, the constituent waves associated with the sun and moon are in phase and produce the so-called *spring tides*. Much smaller *neap tides* are those which occur at the time of the quarter or three-quarter moon when the sun and moon are at 90 degrees to each other as measured from the earth. Their respective following waves are then out of phase and result in smaller tidal ranges (Figure 1.5).

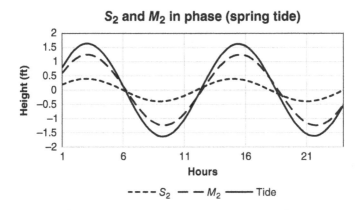

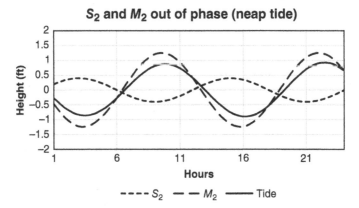

Figure 1.5 The Relationship Between the Semi-diurnal Constituents of the Moon (M_2) and Sun (S_2) for a Spring Tide and a Neap Tide Using the Amplitudes of Those Tidal Constituents for Fernandina Beach, Florida.

The observable daily tidal patterns may be classified into three categories. These are semi-diurnal, mixed, and diurnal. Semi-diurnal tidal patterns (Figure 1.6) have two cycles of nearly equal amplitude during each lunar day of 24.84 hours. Tides along the east coast of the United States are predominately semi-diurnal. Mixed tidal patterns (Figure 1.7) have two cycles which are of unequal amplitude. Tides along the west coast of Florida and the Pacific coast of the United States have predominately mixed tidal patterns. Diurnal tidal patterns (Figure 1.8) have only one tidal cycle per day and are found along the northern and western coast of the Gulf of Mexico and several other places in the world.

In addition to the daily patterns, longer-term observations allow a view of other tidal patterns. For example, a plot of hourly sea level values over a month of observations shows a pronounced pattern with a period of 27.55 days associated with the elliptical orbit of the moon about the earth (Figure 1.9). In such a plot, the spring tides associated with the new and full moons as well as the neap tides associated with the quarter moons may be observed as having greater and smaller ranges.

Similarly, an annual tidal pattern is typically visible as a trend line when the mean tide level (MTL) for each month in the year over a 19-year tidal epoch is plotted (Figure 1.10). That pattern is particularly noticeable in areas such as the northern reaches of the Gulf of Mexico where the tides tend to be considerably lower in the autumn due to prevailing northerly winds.

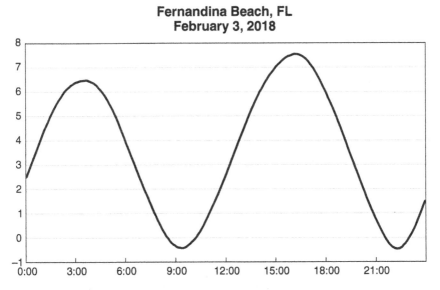

Figure 1.6 Typical Semi-Diurnal Tidal, Pattern Fernandina Beach, Florida.
Source: www.tidesandcurrents.noaa.gov/NOAA/Public Domain.

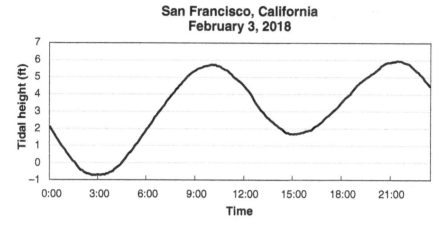

Figure 1.7 Typical Mixed Tidal Pattern, San Francisco, California.
Source: www.tidesandcurrents.noaa.gov/NOAA/Public Domain.

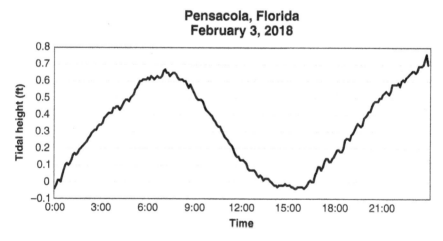

Figure 1.8 Typical Diurnal Tidal Pattern, Pensacola, Florida.
Source: www.tidesandcurrents.noaa.gov/NOAA/Public Domain.

In addition to those described and other relatively short-term constituents associated with astronomic forces, there is another sinusoidal cycle, that associated with the movement of the moon's nodes[1] with a period of 18.6 years, that affects

---

1 Movement of the moon's nodes refers to the movement of the intersection of the moon's orbital plane and the plane of the Earth's equator which completes a 360° circuit in 18.61 years.

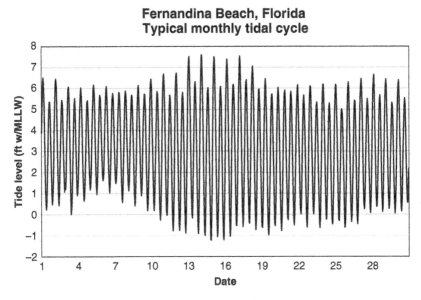

Figure 1.9 Typical Monthly Tidal Pattern – June 2018, Fernandina Beach, Florida, Florida.
Source: www.tidesandcurrents.noaa.gov/NOAA/Public Domain.

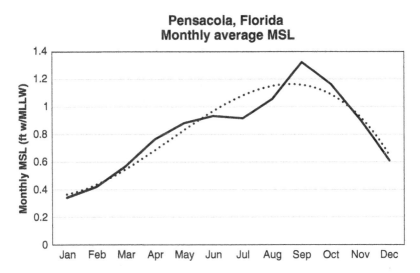

Figure 1.10 Typical Average Monthly Mean Sea Level, Pensacola, Florida – 1999–2018.
Source: www.tidesandcurrents.noaa.gov/NOAA/Public Domain.

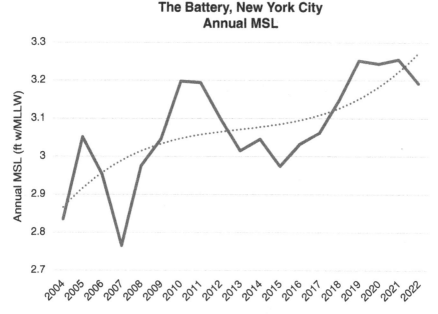

Figure 1.11 18.6 Year Tidal Cycle – 2004:2022, The Battery, New York, NY. In Addition to the Effect of the 18.6 Year Cycle, the Effect of the Upward Long-term Trend May be Seen in this Figure. *Source:* www.tidesandcurrents.noaa.gov/NOAA/Public Domain.

the water level. The effect of that cycle may be seen in Figure 1.11. There is also a long-term cycle with a period of 8.85 years due to the lunar perigee cycle, which has a prominent effect on tides in some areas (Haigh et al. 2011). It is typically apparent as a 4.4-year cycle.

Considering the periodic variations in sea level described earlier, the resultant observed tide is the composite, or algebraic sum, of all the above-mentioned constituent cycles. Although there are theoretically several hundred tidal constituents, not all are significant. Twelve of them are generally considered of primary importance in most locations within the United States (Table 1.1).

One advantage of considering the observed tide as the sum of constituent components is that it allows prediction of tides. With amplitude and phase lag of the constituents determined by tidal observations at a given location and the known constituent periods, the amplitude and time of the tidal extremes may be predicted at a location for years in advance. Lord Kelvin, who developed the tidal harmonic constituent analysis theory in 1867, also designed a tide predicting machine. That device was, in essence, a mechanical computer. It physically summed the amplitudes of the harmonic constituents over a given time and traced the resulting curve. Several similar devices were subsequently constructed

Table 1.1 Principal Tidal Constituents.

Symbol	Period (h)	Description
Semi-diurnal		
M_2	12.42	Principal lunar
S_2	12.00	Principal solar
N_2	12.66	Larger elliptical lunar
K_2	11.97	Lunar – solar semi-diurnal
Diurnal		
K_1	23.93	Luni-solar diurnal
O_1	25.82	Principal lunar diurnal
P_1	24.07	Principal solar diurnal
Q_1	26.87	Larger lunar elliptic
Long period		
M_f	1.000	Radiational
M_m	0.997	Principal lunar – solar
S_{sa}	0.962	Elliptical lunar
S_a	0.929	Second-order lunar

Note that the Subscripts of the Constituent Names Indicate the Frequency of the Cycles.
Source: Modified from Gill and Schultz (2001)/NOAA/Public domain.

and were widely used until well into the second half of the twentieth century. The author has a graphic memory of being introduced, as a newly commissioned officer of the U.S. Coast & Geodetic Survey (now a component of NOAA) to the tide predicting machine being used by that agency in 1961 (Figure 1.12). Tide prediction is now easily accomplished on a desktop computer with far greater precision.

1.2.2 Tidal Datum Planes

A *tidal datum* is a plane of reference based upon average tidal heights. To be statistically significant, all of the periodic variations in tidal height should be included in the average. As a result, a tidal datum is usually considered to be the average of all occurrences of a certain tidal extreme over a *tidal epoch* of 19 years (the period of the longest astronomic cycle affecting the tides, the motion of the moon's nodes with a period of 18.6 years, rounded to the nearest whole year for inclusion of a multiple of the annual cycle associated with the declination of the sun).

Figure 1.12 U.S. Coast and Geodetic Survey's Tide Predicting Machine No. 2. *Source:* NOAA/https://tidesandcurrents.noaa.gov/predma2.html/last accessed 13 November 2023.

Common examples of tidal datum planes include *mean high water* (MHW), defined as the average height of all the high waters occurring over a 19-year tidal epoch, and *mean low water* (MLW), defined as the average of all of the low tides over a 19-year tidal epoch. Other common tidal datums include MTL, which is the plane halfway between MHW and MLW, and *mean sea level* (MSL), which is defined as the average level of the sea over a tidal epoch. The relationship between MSL and MTL varies with location based on the phase and amplitude relationships of the various tidal constituents at those locations. Other commonly used datum planes include *mean higher high water* (MHHW), defined as the average of the higher of the high tides occurring each day over a tidal epoch, and *mean lower low water* (MLLW), which is defined as the average of the lower of the low tides occurring each day over a tidal epoch (Figure 1.13).

Based on the above definitions, it may be seen that determination of a tidal datum involves a relatively simple determination of the arithmetic mean (average) of all the occurrences of a certain tidal extreme over a 19-year tidal epoch. Two less frequently used tidal datum planes which are typically exceptions when used include mean low water springs (MLWS) and equinoctial spring tide. MLWS

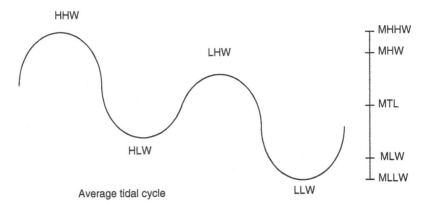

Figure 1.13 Common Tidal Datum Planes.

is the arithmetic mean of the low water heights occurring at the time of spring tides over a tidal epoch. It is usually calculated as a value one half the spring range of tide below MTL and is used in some countries as a hydrographic datum. As discussed earlier in this section, spring tides are those experienced twice a month near the time of the new and full moon when the earth, moon, and sun are in a line. Tidal ranges for spring tides are considerably greater than average tides in most areas due to the gravitational forces of the moon and sun pulling in unison. *Equinoctial spring tides* result twice annually when spring tides occur near the times of the vernal or autumnal solar equinox. At those times, the sun is over the equator, and the paths of the sun and moon are in closest alignment resulting in tidal ranges greater than average spring tides.

It should be noted that the tidal extremes used in calculation of tidal data are not necessarily the highest and lowest water that occur in a tidal cycle. In most areas, especially on the open coast, wind wave action causes a constant and frequent variation in water level. For consistency, since wind wave action is essentially unpredictable, tidal gauges measure the height of stilled water which is approximately half-way between the crest and trough of the wind waves. In areas with sizable wind waves, there can be a significant vertical distance between the crest (or trough) of the wind wave and the stilled water level of high and low water.

1.3   Long-Term Changes in Sea Level

1.3.1   Pre-Historic Trends

Sea level has been in a state of flux throughout the history of the world and continues to be so today. For at least the past half million years or so, the Earth has experienced climate cycles with periods averaging roughly 100,000 years. That period is related to the Milankovitch cycles which are the result of the eccentricity of the Earth's orbit and the tilt and precession of the Earth's axis. Each such cycle included a glacial period during which the earth experienced cooler than average temperatures resulting in the formation of many glaciers and an inter-glacial period with warmer than average temperatures during which many of the glaciers melted. This, in turn, resulted in significant changes in global sea level (Figure 1.14).

It may be seen from Figure 1.14 that the last glacial maximum ended about 15,000 years ago. At that time, the ice began melting and sea level began rising and has generally been on an upward trend since that time. Graphic evidence of the earlier stand of the Gulf of Mexico along the Big Bend area of the Florida Gulf of Mexico coastline may be seen today as a pronounced escarpment, called the Cody Scarp, lying 18 mi or so upland of the current shoreline (Figure 1.15).

As the last glacial period ended, higher temperatures resulted in melting ice-caps, which led to sea level rise, increased precipitation, and greater volume of water in rivers. The scale and rate of that rise in the Gulf of Mexico has been quantified by a report published by the Florida Geological Survey (Basillie and Donaghue 2004). That report provides an analysis of 341 separate radiocarbon-dated sea level indicator points based on 23 independent field studies in the northern Gulf of Mexico. The results of that report, when plotted, provide a good overview of the trend of sea level in the northern Gulf of Mexico over the last

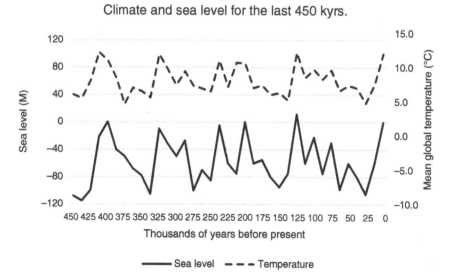

Figure 1.14 Recent Glacial/Inter-Glacial Periods. *Source:* Adapted from Carson (2011).

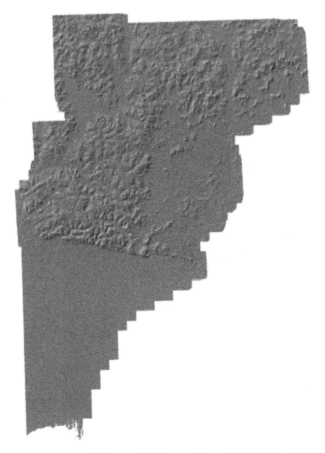

Figure 1.15 Bare-Earth LiDAR Image of Jefferson County, Florida, Showing the Cody Scarp Crossing the Southern Portion of the County.

22,000 years (Figure 1.16). A least squares regression through those points suggests an average rise of 6.0 mm/yr over the last 22,000 years, and 2.3 mm/yr over the last 10,000 years. Those changes are considered to be associated with melting of grounded ice, thermal expansion, and redistribution of water mass, all associated with the long-term climate change that has taken place during the last 25,000 years or so.

The above cited trends reflect averages over thousands of years. Within those periods, there were numerous short-term trends with periods of a few hundred years or so, with sea level trends considerably different from the average over the entire period. As an example, the downward trend in sea level between 12,835 and

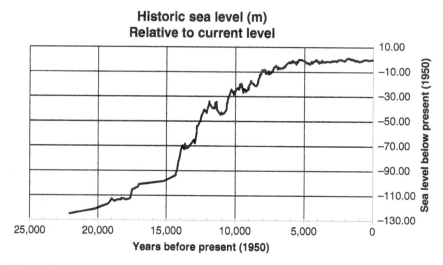

Historic sea level (m)
Relative to current level

Figure 1.16   Estimated Sea Level Change Over Previous 22,000 Years, Based on Data from Basillie and Donaghue (2004), Average Rise Over Last 25,000 years: 6.0 mm/yr, Average Rise Over Last 10,000 years: 2.3 mm/yr.

11,735 before present[2] has been associated with a sudden cooling of the earth believed to be triggered by the impact of an asteroid with earth. That period has been associated with the sudden extinction of the mammoths and a decline of the Clovis culture in North America.

Another period of sea level decline began about 1800 years before present (150 AD) when the sea level was more than a meter above current levels. After that period, the data suggest a falling sea level trend of −3.0 mm/yr for several centuries followed by a generally rising trend. Within the last 1000 years, the data suggest a subperiod of rapid rise between 1000 and 700 years before present (950–1250 AD) during the so-called "medieval warm period" and another of rapid decline during the "little ice age" from 650 to 300 years before present (1300–1650 AD) when much of the world was subjected to cooler winters (Figure 1.17).

Despite the various previously described exceptions, the overall trend in sea level, based on geologic data, has been upward since the last glacial maximum, some 15,000 years ago. The result of that trend has been a gradual retreat of the shoreline. As an example, along the Big Bend area of the Florida's northern gulf coast, the shoreline is believed to be as much as 90 mi landward of its location at the beginning of that retreat. Submerged paleo-channels of rivers that led to the

---

2 For the referenced study, "present" was defined as 1950.

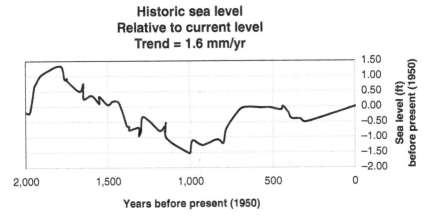

Figure 1.17 Estimated Sea Level Change Over Previous 2000 Years, Based on Data from Basillie and Donaghue (2004), (Enlargement of Portion of Figure 1.16).

shoreline at the beginning of that retreat as well as evidence of pre-historic human settlement are still clearly visible, many miles off the current shoreline (Cole et al. 2017). Therefore, pre-historic geologic data clearly suggest a constantly moving boundary between the oceans and the upland.

1.3.2 Contemporary Trends

In the last 200 years, records from sea level gauges are available that allow direct observation of the rate of sea level changes (Cole 1997), as opposed to reliance on the geologic coastline indicators. Even though more precise that the rates suggested by geologic evidence covered in the last section, the periods covered are much shorter duration. The longest operating of such stations worldwide (Figure 1.18) is in Brest, France, where sea levels have been recorded since 1807. The data from that station indicate a rising trend of 1.1 mm/yr over the entire period of observation although there is an apparent change in trend about 1890. From that point to current, the data suggest a rising trend of 1.6 mm/yr.

The tide station with the longest record of continuous operation in the United States is located in San Francisco, California. It was established by the U.S. Coast & Geodetic Survey in 1854. The records from that station indicate an average rise of 1.96 mm/yr (Figure 1.19) from an apparent datum shift in 1895 through 2018. As may be seen, the data for that station records reflect a relatively constant slope over that period.

Although the sea level change trend indicated by the San Francisco gauge is relatively close to the reported current global rate of sea level rise (1.7 ± 0.5 mm/yr) in

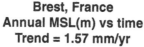

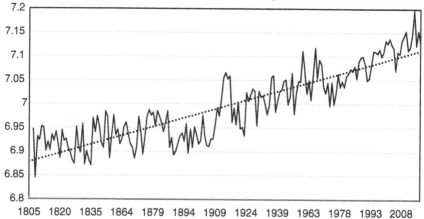

Figure 1.18 Sea Level Trend at Brest, France *Source:* Adapted from www.tidesandcurrents
.noaa.gov.

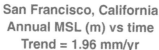

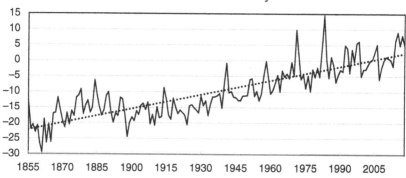

Figure 1.19 Sea Level Trend at San Francisco, California. *Source:* www.tidesandcurrents
.noaa.gov/NOAA/Public Domain.

the twentieth century (NOAA 2009), there is a considerable variation of sea level
trends as recorded in other tide gauges along the U.S. coastline. As an example,
sea level records from a monitoring station at the Battery in New York City, with
records dating back almost as long as the station in San Francisco, show higher
rates of long-term rise (Figure 1.20).

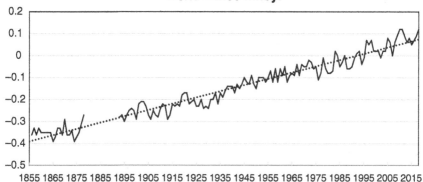

The Battery, New York
Annual MSL (m) vs time
Trend = 2.85 mm/yr

Figure 1.20 Sea Level Trend at the Battery, New York, New York. *Source:* www.tidesandcurrents.noaa.gov/NOAA/Public Domain.

A graphic illustration and possible explanation of the differences in trends at various locations may be seen by examining data from a series of stations along the northern Gulf of Mexico coastline (Table 1.2). By observing gauging data at various locations along that coast, it may be readily seen that the apparent rates of sea level rise appear to vary considerably with location. The high rates appear to reach a peak in areas along the Louisiana coast where there is the highest apparent rise in the United States (e.g. 9.08 mm/yr at Grand Isle). The differing apparent rises are believed to be related to local vertical land movements. Since sea level changes are measured relative to fixed bench marks on land, measured sea level changes include both true sea level changes and vertical land movement due to various factors including earthquakes, tectonic motion, consolidation of coastal sediments, consequences of extraction of oil or water, and responses of the earth to the melting of glaciers (Pugh 2004).

The effect of local vertical land movement on apparent sea level may be graphically illustrated by comparing data for sea level change at Grand Isle, Louisiana, and for Southeast Alaska. At Grand Isle, as may be seen (Figure 1.21), there is an apparent rise in sea level of 9.08 mm/yr, while continuous GPS observations indicate that the ground there is sinking at a rate of 8.1 mm/yr (www .ags.noaa.gov). This suggests that sea level at Grand Isle is actually rising at a rate of only 1.0 mm/yr (9.1–8.1). At Juneau, Alaska, there is a **decline** in apparent MSL of 13.3 mm/yr, while GPS observations there indicate that the ground is rising, apparently due to glacial rebound, at a rate of 15.4 mm/yr. This suggests that sea level at Juneau is actually rising at a rate of only 2.1 mm/yr

Table 1.2 Sea Level Trends along the Northern Gulf Coast.

Station	Rise (mm/yr)
Cedar Key, FL	2.13
Apalachicola, FL	2.38
Panama City, FL	2.43
Pensacola, FL	2.40
Dauphine Is., AL	3.74
Bay Waveland, MS	4.64
Grand Is., LA	9.08
New Canal, LA	5.35
Eugene Is., LA	4.65
Sabine Pass, TX	5.85
Galveston Pier 21, TX	6.51
Galveston Pleasure Pier, TX	6.62
Freeport, TX	4.43
Rockport, TX	5.62
Corpus Christi, TX	4.65
Port Mansfield, TX	3.19
Padre Island, TX	3.48
Port Isabel, TX	4.00

Source: www.tidesandcurrents.noaa.gov/NOAA/Public Domain.

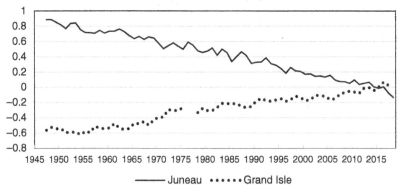

Juneau, AK and Grand Isle, LA Annual MSL (m)

Figure 1.21 Apparent Annual Mean Sea Level, Apparent Average MSL Change: Juneau, AK: −13.3 mm/yr Grand Is., LA: +9.1 mm/yr. *Source:* www.tidesandcurrents.noaa.gov.

(15.4–13.3). This comparison graphically illustrates that in some areas, local geological mechanics are a greater factor in apparent sea level rise than actual sea level.

As may be seen from the previous two sections, sea level change is not a new phenomenon. Rather, sea level has been in a state of flux throughout the history of the world. During the last 15,000 years since the last glacial maximum, evidence suggests that sea level has generally been on a significant upward slope and continues to be so today. Further, that rise has and will continue to have an impact on vegetative and animal life, including humans. Moreover, considering the nature of modern civilization today with its more developed coastal infrastructure, changing sea level will probably have far greater impact today than earlier in the history of the earth even though the current rate is considerably less than in the earlier stage of the current rise.

One important consideration resulting from long-term sea level change is its effect on tidal datum planes. To address this in the United States, tidal datum planes, such as MHW, are calculated using a specific 19-year epoch. Periodically, a new national epoch is adopted by the National Oceanic and Atmospheric Administration (NOAA) after significant change has occurred. At the time of this writing (2015), the National Tidal Datum Epoch is 1983–2001.

1.3.3 Tidal Epochs

As mentioned previously, a tidal datum is defined as an average over a 19-year period known as a tidal epoch. Traditionally, all datum values, published by the National Ocean Service of the NOAA, are referred to a specific time period known as the **National Tidal Datum Epoch**. The policy of National Ocean Service is to consider a new national epoch every 20–25 years to consider adopting a new national epoch. When a new epoch is adopted, adjustments are made to all datum elevations for tide stations published by that agency so that all tidal data throughout the nation are based on a specific time period. The current national epoch, adopted in 2003, is for the 1983–2001 time period.

A modified policy has recently been adopted for regions where the rates of long-term land movement cause anomalously high rates of apparent sea level change. These include areas such as Juneau, Alaska, and Grand Island, Louisiana, shown as examples in the previous section. In areas such as those where the apparent sea level trend exceeds 9.0 mm/yr, a five-year computational period has been adopted to better reflect the current MSL period. Currently, tide data published for those areas are based on the 2012–2016 epoch.

1.4 Shoreline Dynamics

In addition to change in shoreline location due to sea level change, the location of shorelines also changes due to other factors. These include wave and wind action that can wear away, as well as build up the coastline, the action of currents that can erode as well as deposit material, and human activity such dredge and fill activities. The result of all of these forces is that the location of shorelines is constantly subject to change with time. As a result, where the shoreline serves as a boundary, such changes may result in changes in ownership or extent of ownership.

1.5 Variations in Nontidal Waters

Nontidal waters obviously do not demonstrate the predictable periodic variations in water level as those in waters affected by the tides. Nevertheless, variations in the level of such waters do follow certain patterns, generally due to meteorological events. The most obvious of such patterns are those relating to the immediate effect of precipitation. Typically, these are near-vertical rises in water level followed by a return to the long-term trend existing prior to the precipitation event (Figure 1.22). Most nontidal waters also demonstrate changes in water level relating to seasonal meteorological pattern, such as higher levels in the rainy season (Figure 1.23). In addition, most water bodies demonstrate trends due to long-term climate changes or water use (Figure 1.24).

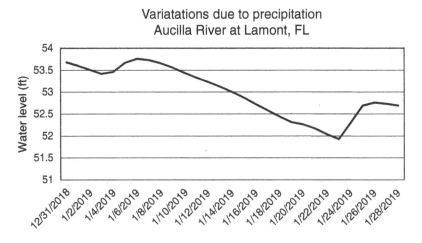

Figure 1.22 Typical Water Level Patterns Associated with Precipitation Events. *Source:* www.tidesandcurrents.noaa.gov.

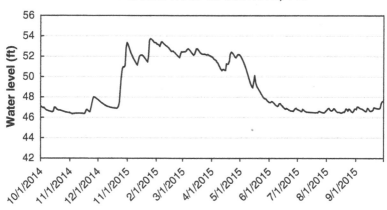

Figure 1.23 Typical Water Level Patterns Associated with Seasonal Precipitation. *Source:* www.tidesandcurrents.noaa.gov.

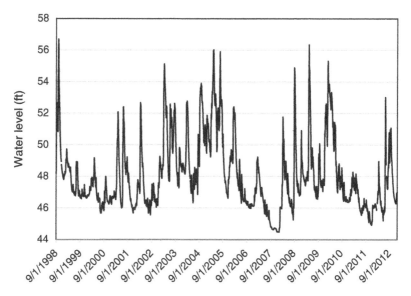

Figure 1.24 Typical Water Level Trend due to Long-Term Effects. *Source:* www.tidesandcurrents.noaa.gov.

References

Basillie, J. and Donaghue, J. (2004). High resolution sea level history for the Gulf of Mexico since the last glacial maximum, Florida Geological Survey Report of Investigations No. 103.

Carson, A. (2011). Ice Sheets and Sea Level in Earth's Past. *Nature Education Knowledge* 3 (10): 3.

Cole, G. (1997). *Water Boundaries*. New York: Wiley.

Cole, G., Hale, J., Ward, D., and Joanas, Z. (2017). Use of Bathymetric LiDAR for Paleo landscape description. Lost and Future Worlds Royal Society Conference, Buckinghamshire, UK.

Gill, S. and Schultz, J. (ed.) (2001). *Tidal Datums and Their Applications*. NOAA, Special Publication NOS CO-OPS.

Haigh, I., Elliot, M., and Pattiaratchi, C. (2011). Global influences of the 18.6 year Nodal cycle of lunar perigee on high tide levels. *Journal of Physical Research* 116 (C6): https://doi.org/10.1029/2010JC006645.

NOAA (2009). *Sea Level Variations in the United States, 1854-2006*. Technical Report NOS Coops 053, Silver Springs, D: NOAA.

Pugh, D. (2004). *Changing Sea Levels*. Cambridge: Cambridge University Press.

2

The Public Trust Doctrine

2.1 Origins of the Public Trust Doctrine

A central concept in this writing is the **Public Trust Doctrine**. That doctrine, which has roots far back in history, considers the sea and other navigable waters as a public common, commonly owned by all mankind. The origin of the doctrine is often attributed to the *Institutes of Justinian*, an early Roman civil code prepared under the direction of Emperor Justinian in 529 AD. Nevertheless, some of that writing is believed to have been based on an even earlier writing, the *Institutes of Gaius*, written about 160 AD by a celebrated early Roman jurist, Gaius.

As an expression of this early doctrine, the *Institutes of Justinian*, stated as follows (Sandars 1874):

> By the law of nature these things are common to mankind – the air, running water, the sea, and consequently the shores of the sea. No one, therefore, is forbidden to approach the sea- shore, provided that he respects habitations, monuments, and buildings, which are not, like the sea, subject only to the law of nations. (Section 1, Book II, Title I)
>
> All rivers and ports are public; hence the rights of fishing in a port or in rivers, are common to all men. (Section 2, Book II, Title I)

A somewhat similar compilation of early Spanish civil law in *Las Siete Partidas,* written and compiled in the thirteenth century under the order of Alfonso the Wise of Castille, more or less tracked and expanded upon the Roman Institutes of Justinian as illustrated by the following (Scott 1931):

> The things which belong in common to the creatures of this world are the following, namely; the air, the rain-water, and the sea and its shores, for every living creature can use each of these things, according as it has need

Sea Levels and Coastal Boundaries, First Edition. George M. Cole.
© 2024 John Wiley & Sons, Inc. Published 2024 by John Wiley & Sons, Inc.

of them. For this reason, every man can use the sea and its shore for fishing or for navigation, and for doing everything there which he thinks may be to his advantage.[1]

The concept of the sea as a public commons has continued through the ages, although with some changes. While the high seas have generally continued to be considered as common to all, various coastal nations have claimed exclusive use of portions of the seas adjacent to their upland for their exclusive use. That earliest concept of a closed territorial sea evolved as a matter of self-defense and self-interest on the part of coastal nations. Such a zone was, for practical purposes, the area which the nation could defend for its exclusive use from hostile shipping. That concept was expressed in a 1703 writing, *De Domino Maris* by Cornelius van Bynkershoek as "Power from land ends where the power of its armament ends." Under that philosophy, the scope of a coastal state's authority over the adjoining waters would extend to the range of its cannons. This was more or less standardized by the so-called "cannon shot" rule which considered a nation's territorial sea to extend a distance from shore that a cannon shot could reach. That distance was generally considered to be 3 mi. That width has varied with time as will be discussed in Chapter 7 (U.S. National and State Maritime Boundaries). Nevertheless, the basic concept of the sea and other navigable waters being commonly owned and distinct from uplands, which are subject to private ownership, has continued to be a prominent component of civil law in most societies.

2.2 Application of the Public Trust Doctrine in the United States

2.2.1 Ownership of Tidal Waters

English common law considered tidally affected waters to be sovereign waters as demonstrated by the following quote (*Attorney General v. Richards*).

> . . . the sea and sea coasts, and as far as the sea flows and reflows, between the high and low-water marks, and all the ports and havens of the kingdom, belong to his Majesty . . .

The United States adopted much of the English common law, with early U.S. case law holding that the marginal territorial seas along the coastline are public trust lands owned in common by the people of each state, since they replaced the

---

1 Partida 3, Title 28, Law 4.

sovereign. This is illustrated in the following 1852 U.S. Supreme Court opinion (*Martin et al. v, Waddell*).

> When the revolution took place the people of each state became themselves sovereign and in that character hold the absolute right to all the navigable waters and the soils under them for their own common use . . .

The question of whether or not such public ownership applies to all tidally affected waters, regardless of their navigability-in-fact, is one that has been the subject of considerable dispute. In some states, public ownership is considered to extend to all submerged lands subject to the ebb and flow of the tide, regardless of actual navigability. Examples are Louisiana, Maryland, Mississippi, New Jersey, New York, and Texas. In those states, all waters subject to tidal ebb and flow are generally considered as sovereign. Yet, in California, Connecticut, Florida, North Carolina, and Washington; the public ownership of tidally affected waters has traditionally been based on the navigability-in-fact of the waters (Maloney and Ausness 1974; Cole 1991).

As one illustration of this distinction, historic Florida case law appears to support the navigability-in-fact approach. For example, one Florida case, *Clement v. Watson,* states that "tidal waters are not under our law regarded as navigable merely because they are affected by the tide." Other cases in that state reflect a similar position. In addition, the navigability-in-fact position is clearly implied in Section 177.28 of the Florida Statutes which states that the ". . . *mean high water line along the shores of land immediately <u>bordering on navigable waters</u> is recognized and declared to be the boundary between the foreshore owned by the state in its sovereign capacity and upland subject to private ownership.*" (emphasis added)

Other states, such as Mississippi and New Jersey, have been generally considered to be "ebb and flow" states. In a challenge of that issue *(Cinque Bambini Partnership v. State of Mississippi)*, both the chancery court and the Mississippi Supreme Court decided in favor of the ebb and flow doctrine. Salient excerpts from the colorful State Supreme court opinion in that case are as follows:

> The early federal cases refer to the trust as including all lands within the ebb and flow of the tide.
> . . . it is our view that as a matter of federal law, the United States granted to this State in 1817 all lands subject to the ebb and flow of the tide and up to the mean high water level, without regard to navigability.
> Yet so long as by unbroken water course – when the level of the waters is at mean high water mark – one may hoist a sail upon a toothpick and without interruption navigate from the navigable channel/area to land, always

afloat, the waters traversed and the lands beneath them are within the inland boundaries we consider the United States set for the properties granted the state in trust.

That Mississippi case was appealed to the U.S. Supreme Court as *Phillips Petroleum Co. v. Mississippi* which concurred with the state court in ruling that the coastal states received all lands over which tidal waters flow, and that the public of Mississippi still owns such land. In the majority opinion, Justice O'Connor wrote that this ruling "will not upset titles in all coastal states [*since it*] does nothing to change ownership rights in states which previously relinquished a public trust claim to tidelands such as those at issue here." (Cole 1991) In the research for this writing, no rulings have yet been found that address the question as to whether or not the ruling has an effect in states which have established navigability-in-fact case law.

2.2.2 Ownership of NonTidal Waters

Early English common law did not consider nontidal waters as sovereign as demonstrated by the following cite from Lord Chief Justice Mathew Hale in his *De Jure Maris* (Hale 1666).

> Fresh rivers of what kind soever, do of common right belong to the owners of the soil adjacent; so that the owners of the one side have, of common right, the property of the soil and consequently the right of fishing, usque filum aquae [to the middle of stream]; and the owners of the other side the right of soil or ownership and fishing unto the filum aquae on their side. And if a man be owner of the land on both sides, in common presumption he is the owner of the whole river, and hath the right of fishing according to the extent of his land in length. . . .

Although subject to private ownership, navigable nontidal rivers were considered by English common law to be public highways as illustrated in the following from Hale's De Jure Maris.

> . . . there be other rivers, as well fresh as salt, whether they flow and reflow or not, are prima facia publici juris, common highways for man or goods or both from one inland town to another. Thus the rivers of Wey, of Severn, of Thames, and divers others, as well above the bridges and ports as below, as well above the flowings of the sea as below, as well where they are become to be of private propriety as in what parts they are of king's propriety, are publick rivers juris publici.

Due to the striking differences in topography between the United States and the British Isles, the U.S. common law regarding nontidal waters differed from that of England as illustrated by the 1876 case of *Barney v. Keokuk* where the U.S. Supreme Court ruled that state ownership in navigable waters extended to inland, nontidal waters as well as tidal waters with the following words:

> The confusion of navigable with tide water, found in the monuments of the common law, long prevailed in this country, notwithstanding the broad differences existing between the extent and topography of the British island and that of the American continent . . . And since this court . . . has declared that the Great Lakes and other navigable waters of the country, are, in the strictest sense, entitled to the denomination of navigable waters, and amenable to the admiralty jurisdiction, there seems to be no sound reason for adhering to the old rule as to the proprietorship of the beds and shores of such waters. It properly belongs to the States by their inherent sovereignty . . .

Thus, some states, in their sovereign capacity, claim title to the beds under navigable waters whether or not the waters are tidally affected. Even in the states that do not claim public ownership of such waters, navigable nontidal waters are generally still considered to be public highways and subject to a public easement.

In nontidal waters, whether or not the waters are part of the public trust generally is a question of navigability-in-fact. As an example of how such waters are defined, case law in Florida *(Odom v. Deltona)* offers clarification to that state's definition in nontidal waters. In that case, the court held that *"Florida's test for navigability is similar, if not identical, to the Federal Title Test."* Understanding of the federal test may be obtained from the following (*The Daniel Ball*):

> . . . and they are navigable in fact when they are used or susceptible of being used in their ordinary condition, as highways for commerce, over which trade and travel are or may be conducted in the customary modes of trade and travel over water.

Thus, the definition of which waters are part of the public trust may vary with whether or not the water is tidally affected and whether or not it is navigable. Further, as discussed in the following chapters, definitions for such boundaries often vary with the jurisdiction in which the property is located.

2.2.3 Application to Adjoining Non-Navigable Areas

In many situations, there are adjoining non-navigable waters, such as coves and tributary streams, adjacent to navigable waters. In such cases, questions sometimes arise as to whether the submerged land under such waters is part of the navigable water body or a separate water body requiring its own determination of navigability. Obviously, this is not an issue with tidal waters in those jurisdictions recognizing tidal influence as the criteria for sovereign ownership. Yet, this can be a serious issue in other jurisdictions.

A typical situation in which this becomes an issue involves a non-navigable indentation adjoining a larger navigable water body. In such a situation, the submerged lands associated with the larger water body are considered as sovereign unless conveyed by the state. This would include the shallow, nearshore waters of the larger water body. The question at issue is whether the waters in the adjoining non-navigable water is also sovereign and, if not, where the dividing line is located.

If the indentation is a non-navigable stream, as opposed to a bay, it would seem logical to apply the technique described in Chapter 5 under international law to separate rivers from the sea. That process involves a headland-to-headland line. If the indenture is not a stream, a logical approach is to also apply international law standards, as described in Chapter 5, to determine whether it part of the primary water body or a separate entity requiring its own determination of navigability.

In Florida, this approach has been substantiated by the State Supreme Court (*Clement v. Watson*). In that case, the entrance to a cove was blocked by a berm over which the only navigable passage was a man-made channel. In another Florida case on this topic (Baker v. State ex rel Jones), the court held that a non-navigable arm of a navigable lake was not sovereign and thus not subject to public use.

2.2.4 Floodplains Adjoining Navigable Waters

In some riverine systems, especially in regions with low relief, there are wide, low-lying floodplains paralleling the navigable channel of the river. Typically in such situations, there is a natural levee between the channel of the river and the floodplain. In some areas, the floodplain contains heavily vegetated swamps inundated during the flood stages of the river. Some areas of the floodplain are lower than the levee. Water from the flood stages of the river, upland runoff, or water that has entered through breaks in the levee is often entrapped in such areas for considerable portions of the year.

At times, questions arise as to whether low-lying portions, or even the entire floodplain, are considered sovereign land. In some states, such areas have been the subject of claims of public ownership. The rational for such claims is that such areas may be lower than the ordinary high water line; and that although evidence of the ordinary high water line may be found at the bank of the river, additional evidence of a similar nature may be found at the upland edge of the floodplain.

Despite such claims, case law appears to clearly require the exclusion of the flood-plain from the sovereign bed of the water body. Examples of this are as follows:

> It neither takes in overflowed lands beyond the bank, nor includes swamps or low grounds liable to be overflowed but reclaimable for meadows or agriculture. . . (Borough of Ford City v. United States; Howard v. Ingersoll).
>   . . . nor the line reached by the water at flood stages (State ex. rel. O'Conner v. Sorenson).
>   It is the land upon which the waters have visibly asserted their domin-ion, and does not extend to nor include that upon which grasses, shrubs and trees grow, though covered by the great annual rises. (Harrison v. Fite).
>   The high water mark on fresh water rivers is not the highest point to which the stream rises in times of freshets . . . (Dow v. Electric Company; Tilden v. Smith).

Further clarification as to the floodplain question may be obtained from the 1849 and 1850 Acts of Congress that granted swamp and overflowed lands to many of the states. Swamp lands are defined as lands that "require drainage to fit them for cultivation. Overflowed lands are those which are subject to such peri-odical or frequent overflows as to require levees or embankments to keep out the water and render them suitable for cultivation." (*San Francisco Savings Union v. Irwin*). "Overflowed lands include essentially the lower level within a stream flood plain as distinguished from the higher levels. . ." (Bureau of Land Management 1973). Based on these definitions, the floodplain appears to fall into the classifica-tion of swamp and overflowed land rather than sovereign land.

Examination of the original Government Land Office (GLO) surveys further substantiates the exclusion of the floodplain from sovereign land. In most cases, the original surveyors meandered along the limits of the bed and not the edge of the floodplain. Earlier guidelines to the GLO surveyors, as well as current instruc-tions, direct such a location. For example, the 1968 Manual of Government Surveying (Hawes 1868) addresses this issue as follows:

> . . . care must be taken in times of high water, not to mistake the margins of bayous or the borders of overflowed marshes or "bottoms" for the true river bank.

References

Bureau of Land Management, *Manual of Instruction for the Survey of the Public Lands of the United States,* 1973.

Cole, G.M. (1991). Tidal water boundaries. *Stetson Law Review* XX (1–2): Fall 1990 & Spring 1991.

Hale, M. (1666). De Jure Maris. Reprinted in S. Moore, *A History of the Foreshore and the Law Relating Thereto*, 3rd ed. 1888.

Hawes, J.H. (1977). Manual of United States Surveying, Reprinted by Carben Surveying Reprints; originally published by J.B. Lippincott, 1868.

Maloney, F.E. and Ausness, R.C. (1974). The use and legal significance of the mean high water line in coastal boundary mapping. *The North Carolina Law Review* (December).

Sandars, T.C. (1874). *The Institutes of Justinian.* London: Longmans, Green & Co.

Scott, S.B. (1931). *Las Siete Partidas.* New York: Commerce Clearing House.

Case Law Cites

Attorney-General v. Richards (2 Anstr. 603)

Borough of Ford City v. United States, 345 F. 2d. 645 (1965).

Cinque Bambini Partnership et al. v. State of Mississippi et al., Miss, 491 So. 2d 508 (1986).

Clement v. Watson, Fla.,58 So. 25 (1912).

Dow v. Electric Company, N.H., 45 A 350 (1899).

Harrison v. Fite, 148 F 781 (1906).

Howard v. Ingersoll, 54 U.S. 381, 427 (1851).

Martin v. Waddell, 41 U.S. (16 Pet.) 367 (1842).

Odom v. Deltona, Fla., 341 So. 2d. 977 (1976).

State ex rel O'Conner v. Sorrenson et al., 271 N.W. 234,222 Iowa 1248 (1937).

Tilden v. Smith, Fla., 113 So. 708 (1927).

3

Boundaries in Public Trust Tidal Waters

Under the Submerged Lands Act (*43 U.S.C., s1301-1 1970*), coastal waters adjacent to each of the coastal states are considered to be in the ownership of the public of those states. Further, all natural interior tidal waters within each state's boundaries are considered to have been conveyed to the public of each of the U.S. coastal state with statehood. Three approaches to the definition of the boundaries of those tidal public trust waters and the bordering uplands, subject to private ownership, are prevalent in the United States. Those approaches include the Anglo-American Common Law, the North-Atlantic Low Water States, and the Civil Law.

3.1 Boundary Definitions in Various Jurisdictions

3.1.1 Anglo-American Common Law

The Anglo-American Common Law has been adopted by most of the U.S. coastal states in defining the upland boundaries of tidally affected public trust water. For those states, that boundary is considered to be the **mean high water line**, which represents the average upper reach of the daily tides. The accepted definition of that line is that formed by the intersection of the rising coastline and the average of the tidal high waters over a period defined by the astronomical forces affecting the tides. Thus, it uses the average of observed tidal high waters over a complete tidal epoch of 19 years or the equivalent based on a correlation using short-term simultaneous observations with a control station with established 19-year mean values.

Possibly the earliest definition of the Anglo-American Common Law definition of a coastal boundary was provided in the late 1400s by Thomas Diggs, a surveyor, engineer, and lawyer, in a book entitled "*Proofs of the Queen's Interest in Land Left by the Sea and the Salt Shores Thereof*." That approach of using the tides to define

Sea Levels and Coastal Boundaries, First Edition. George M. Cole.
© 2024 John Wiley & Sons, Inc. Published 2024 by John Wiley & Sons, Inc.

the boundary was expanded a century or so later in a book by Lord Chief Justice Matthew Hale. Hale's *De Jure Maris* stated that the foreshore, the area overflowed by "ordinary tides or neap tides which happen between the full and change of the moon," belongs to the crown. A latter British case, Attorney General v. Chambers[1] ruled that the boundary was defined by "*the average of medium tides in each quarter of a lunar evolution during the year*."

The modern definition of that boundary for most of the United States was first stated by the Supreme Court in the landmark case of *Borax Consolidated Ltd. v. City of Los Angeles*[2] as follows:

> In view of the definition of the mean high tide . . . and the further observation that . . . there should be a periodic variation in the rise of water above sea level having a period of 18.6 years . . . in order to ascertain the mean high tide line with requisite certainty . . . an average of 18.6 years should be determined as near as possible.

The basis of the 18.6-year period mentioned in the Borax decision is that which includes at least one occurrence of all of the astronomical cycles affecting the daily tides. These include cycles associated with the daily orbits of the moon about the earth, the rotation of the earth in relation to the sun, the monthly lunar and annual solar declination cycles, and the 18.6-year cycle for the regression of the moon's nodes (see Chapter 1). In practice today, a 19-year averaging period, rather than 18.6 years, is used for inclusion of an integral number of the annual cycles associated with the declination of the sun.

As may be seen from the definition as it evolved through the British common law and U.S. case law, the boundary, currently called the mean high water line, represents an attempt to define the average upper reach of the daily tide as the boundary between publicly owned submerged lands and bordering uplands subject to private ownership. Since that lines vary from day to day, this definition defines the *average* daily upper reach of the tides as the boundary. The result is in a line which is exceeded by the high tide by approximately one-half of the tidal cycles.

Most of the U.S. coastal states have followed the English common law and its evolved definition as defined in the Borax decision. Alabama, Alaska, California, Connecticut, Florida, Georgia, New Hampshire, Maryland, Mississippi, New Jersey, New York, North Carolina, Oregon, Rhode Island, South Carolina, Texas, and Washington have followed that course (Maloney and Ausness 1974; Cole 1977). Some states have codified their common law on this subject. As an

---

1 *A17* Eng. Rul. Cas.555,1889.

2 *Borax Consolidated Ltd. v. City of Los Angeles,* 296 U.S. 10, 1935.

example, in Florida, the Coastal Mapping Act of 1974 (Chapter 177, Part II, Florida Statutes) declares that "the mean high water line along the shores of lands immediately bordering on navigable waters is recognized and declared to be the boundary between the foreshore owned by the State in its sovereign capacity and upland subject to private ownership." That statute defines the mean high water line using the concept stated in the Borax case as follows:

> Mean high water means the average height of the high waters over a 19-year period. For shorter periods of observation, mean high water means the average height of the high waters after corrections are applied to eliminate known variations and to reduce the result to the equivalent of a mean 19-year value. (Section 177.27 (14), Florida Statutes).

3.1.2 North-Atlantic Low Water States

Five States along the North-Atlantic coastline of the United States (Delaware, Maine, Massachusetts, Pennsylvania, and Virginia) have adopted *mean low water* as their coastal boundary. Mean low water is defined as the average of all of the tidal low waters over a 19-year tidal epoch. In the New England states, this policy is based on a Massachusetts colonial ordinance of 1641–1647 as follows:

> . . . in all creeks, coves and other places, about and upon salt water where the Sea ebs and flows, the Proprietor of the land adjoining shall have proprietie to the low water mark where the Sea doth not ebb above a hundred rods,[3] and not more wheresoever it ebs farther.

Based on the language of that ordinance, the waterward boundary of lands bordering on tidal waters in those jurisdictions is the mean low water line unless that line is 100 rods distant from the mean high water line. If there is a 100 rod or greater distance between the two lines, then the boundary is a line 100 rods seaward of the mean high water line. With that definition, ownership of riparian or littoral lands under those legal systems includes the tidal flats often found in such areas. An early Massachusetts supreme court decision, *Storer v. Freeman*, provided an explanation for this approach: "For the purposes of commerce, wharves erected below high water mark were necessary. But the colony was not able to build them at public expense. To induce persons to erect them, the common law of England was altered by an ordinance, providing that the proprietor of land adjoining on the sea or salt water, shall hold to low water mark, where the tide does not ebb more than one hundred rods, but not where the tide is a greater distance."

---

3 A rod is unit of distance measure representing one-quarter of the length of a surveying chain or 16½ ft.

Even though the ownership line is the mean low water line, states adopting this approach for their coastal boundaries typically reserve a public easement up to the mean high water line even though public ownership extends only to the mean low water line. Such a reservation is illustrated by the following (*Tinicum Fishing Co. v. Carter*):

> The title of a riparian owner extends to low-water mark, not absolutely in tidal streams, but subject to the public right of passage when the tide is high.

3.1.3 Civil Law

The civil law definition for coastal boundaries prevails in areas where the land title has its roots in a grant from a sovereign power where civil law, such as the previously mentioned Roman Institutes of Justinian, prevailed. The distinction between the Anglo/American common law and the civil law is clearly distinguished in the opinion for the landmark case of Borax Consolidated v City of Los Angeles with the following statement: *By the civil law, the shore extends as far as the highest waves reach in winter. But by the common law, the shore* "is confined to the flux and reflux of the sea at ordinary tides." That cite is based on the definition of the boundary from the Institutes of Justinian (Sandars 1874) which stated that:

> The sea-shore, that is, the shore as far as the waves go at furthest, was considered to belong to all men . . . . . . The sea shore extends as far as the greatest winter floods runs up.

That definition is best understood when it is realized that the Roman civil law code evolved in an area in the Mediterranean with little daily tidal range (as opposed to the greater daily tidal range in the British Isles). As a result, that law defines the coastal boundary in terms of seasonal water levels rather than in terms of daily tide levels.

As an example of the use of the civil law definition, the State of Louisiana has laws that hold that the waters of the Gulf of Mexico, including the shores of the Gulf, are in public ownership and are considered to extend to the "highest tide during the winter season" (Article 451. LA Civil Code and LA Rev. Stat. 49:3). Interestingly, the high tide during the winter in the northern Gulf of Mexico is usually *less* than the high tide during other parts of the year because of the prevailing northern winds in that area during winter. The State of Hawaii also appears to follow the civil law approach with a coastal boundary defined as "the upper reach of the wash of the waves" (Maloney and Ausness 1974).

Another example is the State of Texas which has recognized the civil law definition in areas of the state with origins of land title in Spanish or Mexican land grants. The Commonwealth of Puerto Rico also follows the civil law. Interestingly, for both Texas and Puerto Rico, it has been held that the limit of ownership is controlled by old Spanish civil law contained in *Las Siete Partidas* written in the thirteenth century, as opposed to the Roman civil law. In the Spanish civil law, the boundary is defined as follows:

> . . . e todo aquel lugar es llamado ribera de la mar quanto se cubre el agua della, quanto mas crece en todo el año, quier en tiempo del invierno o verano . . . (Partida 3, Title 28, Law 4)

The correct interpretation of the above clause has been the subject of considerable judicial debate. Possibly the leading English translation of these laws (Scott 1931) translates the context as follows with the underlined text representing the Spanish clause provided above.

> The things which belong in common to the creatures of this world are the following, namely; the air, the rain-water, and the sea and its shores, for every living creature can use these things, according as it has need of them. (Partida 3, Title 28, Law 3)
>
> Every man can build a house or a hut on the sea shore which he can use whenever he wishes; and any person can erect there another edifice for his own benefit, provided the common custom of the people is not violated; and he can construct galleys and any other kind of ships and dry nets there and make new ones if he desires to do so; and so long as he is working there or is present no one else should molest him so that he cannot use and be benefited by all these things, and by others like them, in the manner aforesaid; and all that ground is designated the shore of the sea which is covered with the water of the latter at high tide during the whole year, whether in winter or in summer. (Partida 3, Title 28, Law 4)

It may be readily seen that the Spanish civil law differs from the Roman in that rather than use of the phrase "*as far as the highest waves reach in winter*" as in the Roman Institutes of Justinian; this Scott interpretation of *Las Siete Partidas* calls for the line where the land is covered by the high tide whether in winter or summer (*quier en tiempo del invierno o verano*). Use of the words suggesting "high tide" calls for a boundary determined by use of tidal data as opposed to the use of water levels caused by metrological events. That interpretation is credible since, unlike the Mediterranean coast of Italy (home of the Roman law) with an almost imperceptible tidal range, Spain has coastlines on both the Mediterranean Sea and Atlantic Ocean. Therefore, the authors of *Las Siete Partidas* would have been far more aware of the prominent daily tidal cycle and its effect on the shore than the authors of the Roman *Institutes of Justinian*. Yet, considering that this section of the code also states that the public commons includes the shore where fishermen have the right to dry their nets and construct huts, this suggests a boundary considerably higher than the mean high water line of the Anglo-American common law.

One judicial interpretation of *Las Siete Partidas* was made by the Supreme Court of Texas in *Luttes v. State in 1950*. In that trial, testimony provided various interpretive translations which suggest that the proper meaning is alternatively the highest swell of the year, the highest tide of the year, or an average high tide. Much of the deliberation centered about the historic meaning of the Spanish verb "*crecer*" or "to grow, expand or rise up" and whether it means "rise of the water of the sea by effect of the tide (*marea*)." The Court concluded that the language of *Las Siete Partidas* implied an average tide and that the applicable rule is that of the "average of highest daily water computed over or corrected to the regular tidal cycle of 18.6 years. This means in substance mean high tide." That language was clarified in a response to a motion for rehearing with the statement that it was the court's intention that the line was actually that of "mean higher high tide,[4] as distinguished from the mean high tide of the Anglo-American law." That clarification was apparently in recognition that the prevailing tide along the South Texas coast is diurnal, with only one tidal cycle per day most of the month so that the average of the daily high tides is essentially the mean higher high water (MHHW), which is the average of the highest of tides that occur each day.

A considerable different interpretation prevails in Puerto Rico, which also follows the civil law as prescribed in *Las Siete Partidas*. The decision in case of *Rubert Armstrong v. E.L.A.*[5] held that Puerto Rico has

> el mismo criterio de la ley de Costas de España, de que el límite de la zona marítimo-terrestre será el que abarque un mayor ámbito físico, hasta donde alcancen las olas en los mayores temporales o, cuando lo supere, la línea de pleamar máxima viva equinoccial.

---

4 Mean higher high water is the average of the higher of the two daily tidal high waters each day over a 19-year tidal epoch. (See Section 1.1.2 of this writing.)

5 97 D.P.R. 588 (1969), el Tribunal Supremo de Puerto Rico.

This may be loosely translated as having "the same criteria as the coastal law of Spain, with the coastal boundary including the greatest physical environment, where the waves of the greatest storms reach, or when it is higher, the line of the equinoctial spring tide.[6]"

Furthermore, territorial guidelines for determining coastal boundaries in Puerto Rico[7] state:

> Para determinar el alcance tierra adentro del criterio mareal se pueden utilizar los indicadores bióticos o el alcance de la pleamar máxima viva equinoccial . . . .

which may be loosely translated as "for determining the landward reach of the tides, you can use the biological indicators **or** the reach of the **equinoctial spring tide**" (emphasis added).[8]

Another approach that has been suggested in Puerto Rico (Cole 2011) is the use of the *highest astronomical tide*, which is the elevation of the highest predicted astronomical tide expected to occur at a specific location over a tidal epoch of 19 years. This occurs when the major astronomic cycles which cause the observable daily tide are all "in phase" and at their maximum. This latter datum would more or less represent the most landward tide possible under average weather conditions and can be determined with a knowledge of the harmonic constituents of the tide at a specific location.

All of the alternatives discussed above for compliance with the original intention of the Spanish code of *Las Siete Partidas* offer the objectivity and mathematical certainty of tidal boundaries. Yet, it would seem that the interpretation offered by the Luttes case in Texas does not meet the definition of including both the sea and the shore as part of the public commons, while either the equinoctial spring tide or the highest astronomical tide does include at least a portion of the shore and thus best meets the criteria stated in the Spanish code.

---

6 Spring tides are those experienced twice a month near the time of the new and full moon when the earth, moon, and sun are in a line. Tidal ranges for spring tides are considerably greater than average tides in most areas due to the gravitational forces of the moon and sun pulling in unison. Equinoctial spring tides result twice annually when spring tides occur near the times of the vernal or autumnal solar equinox. At those times, the sun is over the equator and the paths of the sun and moon are in closest alignment resulting in tidal ranges greater than average spring tides. (See Section 1.1.2 of this writing.)

7 Section 3.2.1.a (2), Manual de Procedimientos para el Deslinde del Limite Interior Tierra Adentro de los Bienes de Dominio Publico Maritimo Terrestre, DRNA, 1999.

8 In practice in Puerto Rico, the boundary is usually determined by vegetative and geological features reflecting highest water levels, including storm surges. Tests of such a line show that it is higher than either the equinoctial spring tide or the highest astronomic tide and therefore reflects a line created by higher water levels such as those associated with storm events (Cole 2011).

3.2 Local Tidal Data Determination

3.2.1 Local Variation

Because of numerous local topographic forces shaping the tide, a tidal datum is very much a local phenomenon. In the oceans and other large water bodies, there is direct response to the tide producing forces with the pattern and magnitude of such oscillations governed by the volume of water, the natural oscillation period of the ocean basin, and the topography of the basin. As an example of the effect of topography, the sea surface has been noted to be depressed as much as 192 ft over ocean trenches and may bulge as much as 16 ft over seamounts (Duxbury 1989).

As a tide wave approaches a land mass, the shoaling of the continental shelf tends to increase the amplitude, very much like a tsunami or storm surge wave. Then, as the wave enters into openings in the coastline, the tide waves are further shaped by local topography. For many estuaries, the amplitude of the tidal wave is gradually diminished. For others, such as the Bay of Fundy in southeastern Canada where the average tidal range is 7 ft at the entrance, the amplitude is continually increased to almost 36 ft at the head of the Bay. In other estuaries, the maximum tidal range is at locations other that at the head of the estuary. This is due to the fact that in those estuaries, the tide wave height at any given time is a product of both the primary wave originating in the open sea and moving up the estuary and a reflected wave originating at a barrier or partial barrier, often at the head of the estuary (Cole 1994). The resultant wave height at a given location is the sum of the two waves. At locations in the estuary where the direct and reflected waves are "in phase," the highest portion of both waves occur at the same time as do the lowest portions of both waves. This results in a relatively large observed tidal range at those points and a smaller tidal range where the two waves are out of phase.

Another factor influencing tidal range within estuaries is the Coriolis effect. In wider estuaries, greater tidal ranges are experienced in areas closer to the right-hand side of the estuary when traveling away from the ocean. Such variation is due to the Coriolis effect which causes a right-hand veer to ocean currents in the Northern Hemisphere (Williams 1962). For the incoming tidal wave, that veer results in a piling up of water on the right side of the estuary (when traveling away from the ocean), which creates a slope in the water level across the channel with the greater high tide experienced on the right side. On the outgoing tide, a similar right-hand veer to the water, now flowing in the opposite direction, results in an opposite slope and a lower low tide on the right side when traveling away from the ocean (Defant 1958).

The effect of both the Coriolis effect and a reflected tidal wave may be observed in The Chesapeake Bay where a plot of the tidal range variation going up the estuary has a sinusoidal pattern which is typical of linear estuaries (Figure 3.1). In the Bay, differences of as much as a foot in tidal range exist between the eastern and western shore with the larger range of the eastern shore as the result of the right-hand side as the tidal wave travels away from the ocean and up the Bay. Also, in the Chesapeake, the abrupt end of the Bay just north of Harve de Grace, Maryland causes a reflective wave that travels back down the Bay. When the incoming and reflected wave are "in phase," such as just north of the mouth of the Potomac River, the range is considerably greater. Likewise, when the two waves are completely out of phase, such as just North of Baltimore and at the 50 mi distance, the resultant tidal range is considerably reduced (Cole 1994).

As the result of such local variation, it is clearly important that a tidal datum used for a boundary determination is the one that has been established in the immediate area of its use. This may be accomplished by use of data from an existing tide station such as those created by NOAA along the nation's coastline at which tidal datums based on the current tidal epoch have been established. If such a station does not exist in the immediate vicinity of the need, local tides may be observed simultaneously with observations at such an established tide station to allow calculation of a local datum.

In some cases, instead of establishing a new tide station, it may be possible to establish the elevation of a datum at the needed location by interpolation between existing tide stations. Such an approach is appropriate on stretches of unbroken open ocean coast lines where tidal range variation generally is relatively linear. Interpretation is not recommended where there are intervening inlets, where either existing tide station is near an inlet, where the coastline is irregular, or where the distance between the two stations is excessive. Linear interpolation is also not recommended within estuaries since tidal datum variation is generally nonlinear in such waters. Research (Cole 1994) has found that the variation pattern within estuaries can often be described by a high-order polynomial equation, but this varies with the configuration of the estuary. As a result, it is recommended that a new tidal station be established when a tidal datum is needed in such waters.

Depending on the distance and hydrographic conditions between the site and the control station, the recommended duration of such observations may vary from a few days for a nearby control station to a year or more for situations where the control station is a considerable hydrographic distance away. The observations may then be correlated using a simultaneous comparison process (Cole 1997).

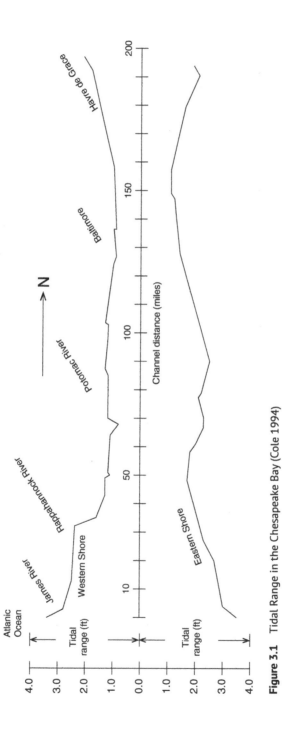

Figure 3.1 Tidal Range in the Chesapeake Bay (Cole 1994)

3.2.2 Tidal Data Calculations

As discussed in Chapter 1, the primary determination of a tidal datum involves the relatively simple calculation of the arithmetic mean, or average, of all the occurrences of a tidal extreme over a 19-year tidal epoch. For many applications, such a primary calculation is obviously not practically considered as the need for a local datum. When this is the case, a local datum may be established by short-term simultaneous observations of the tide at the local site and at a nearby existing tide station with known 19-year mean values. The standard method, known as the **range-ratio method,** for accomplishing this follows (Marmer 1951; Cole 1997):

3.2.2.1 Standard (Range Ratio) Method of Simultaneous Comparison

This method requires knowledge of the 19-year averages for the mean range (MR_c) and mean tide level (MTL_c) at the control station as well as the observational period averages of the range (R_c and R_s) and averages of the observed and low tide (TL_c and TL_s) at both the control and subordinate stations. Note that the control station values are designated by a subscript "c" and the subordinate station values with a subscript "s".

The first step is to calculate the equivalent 19-year mean range (MR_s) at the subordinate station with the following equation:

$$\frac{MR_s}{R_s} = \frac{MR_c}{R_c} \quad \text{which may be restated as} \quad MR_s = MR_c \frac{R_s}{R_c} \tag{3.1}$$

The second step is to calculate the equivalent 19-year mean tide level (MTL_s) at the subordinate station with the following equation:

$$MTL_s - TL_s = MTL_c - TL_c \quad \text{which may be restated as} \quad MTL_s = MTL_c - TL_c + TL_s \tag{3.2}$$

Then, the equivalent values for the 19-year mean high water and mean low water at the subordinate stations may be calculated as follows:

$$MHW_s = MTL_s + \frac{MR_s}{2} \tag{3.3}$$

$$MLW_s = MTL_s - \frac{MR_s}{2} \tag{3.4}$$

If the value for MHHW and mean lower low water (MLLW) at the subordinate station is desired, those values may be calculated with a knowledge of the observational period averages of the high water and higher high water (HW, HHW) and low water (LW) and lower low water (LLW) at both the control and subordinate stations.

$$MHHW_s = MHW_s + \frac{HHW_s - HW_s}{HHW_c - HW_c}\left(MHHW_c - MHW_c\right) \tag{3.5}$$

$$MLLW_s = MLW_s - \frac{LW_s - LLW_s}{LW_c - LLW_c}\left(MLW_c - MLLW_c\right) \tag{3.6}$$

The standard method was developed for use with averages taken over a month or year simultaneously at the control and subordinate station. Yet, where the control station is relatively close to the location where a new datum is needed, satisfactory results may usually be obtained by observing only a few tidal cycles simultaneously. Often, the most efficient method for such observations involves manually read staffs since only an hour or two of observations on both sides of both the high and low tidal extremes at both stations allows a calculation.

Amplitude Ratio Method of Simultaneous Comparison – The standard method requires observation of both the high and low tidal extremes at the control and subordinate stations during the period of simultaneous comparison. If values for a mean high water datum are needed in areas where only the upper portion of most tidal cycles are observable (such as marshes, mud flats, and tidal creeks with wide, flat intertidal zones), an alternate method, the amplitude ratio method may be used.

The Amplitude Ratio Method (Cole 1981) – The amplitude ratio method was derived to mathematically duplicate the results of the standard method, based on the assumption that the observed tide wave is in the form of a perfect sine wave. When used properly where both the control station and subordinate station are relatively close with similar tidal characteristics and if the wave is not substantially distorted from a sine wave shape, the results of this method should closely approximate those which would have been achieved with the standard method had a complete tidal cycle been observed.

The amplitude ratio method is a graphic process and is difficult to explain clearly in text. Basically, the method is based on the fact that with two sine waves of equal wave length, the ratio of the tidal ranges of two tidal cycles is proportional to the ratio of vertical distances between the peaks of the tidal cycles and a horizontal line which intersects the two tidal curves with the same differences in time (Figure 3.2).

The process involves selection of a time interval (T) and the determination of the height of where that interval intersects the curve of the control and subordinate tide curves for the same cycle. The difference in elevation between the peaks of the curves and the elevations at which the time interval intersects the tide curves (A_c and A_s) is proportional to the respective ranges of the two tidal cycles. Thus, the range of the subordinate tidal cycle, had it been observable, may be calculated as the product of the observed range at the control station and the ratio

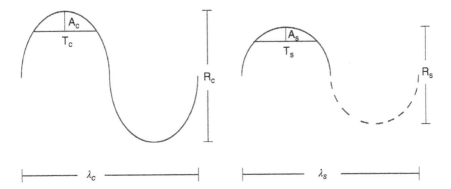

Figure 3.2 Amplitude Ratio Method (Cole 1981)

of the differences in elevation (A_c and A_s) between the peaks of the curves and the elevations at which the time interval intersects the tide curves (Equation 3.7).

$$R_s = R_c \frac{\left(A_s\right)}{A_c} \tag{3.7}$$

Then the equivalent value for average observed tide level (TL_s) may be calculated (Equation 3.8).

$$TL_s = HW_s - \frac{R_s}{2} \tag{3.8}$$

Once the equivalent of the observed range and the average observed tide level at the subordinate station is determined, the mean range (MR_s), mean tide level (MTL_s), mean high (MHW_s), and mean low water (MLW_s) may then be calculated by Equations 3.1 through 3.4 as with the standard method.

Tidal datum elevations for the less frequently used and more complex datum planes, such as equinoctial spring tide or highest astronomic tide, are typically established using predicted tides. Otherwise, unusual weather conditions may result in a considerable skew of the results.

3.2.2.2 Example of Tidal Data Calculation (Standard Method)

To illustrate typical calculations of tidal data, the use of the standard (range ratio) method data from two actual tide stations for a single tidal cycle will be used (Table 3.1 and Figure 3.3). The provided data are intentionally unitless.

From the tabulated data in Table 3.1, it may be seen that control station high water extreme (HW_c) is 2.93 and that the control low water extreme (LW_c) is 1.49. It may also be seen that the subordinate high water (HW_s) is 1.67, and the subordinate low water (LW_s) is 0.22. This results in the observed range at the control

Table 3.1 Typical Tidal Data

Time	Control	Sub.	Time	Control	Sub.	Time	Control	Sub.
6:00	1.14	0.59	9:12	2.82	1.57	12:24	2.15	0.26
6:06	1.20	0.66	9:18	2.84	1.54	12:30	2.11	0.26
6:12	1.27	0.74	9:24	2.88	1.52	12:36	2.08	0.24
6:18	1.35	0.81	9:30	2.89	1.50	12:42	2.04	0.23
6:24	1.41	0.89	9:36	2.91	1.48	12:48	1.99	0.23
6:30	1.48	0.97	9:42	2.92	1.46	12:54	1.95	0.23
6:36	1.55	1.08	9:48	2.92	1.44	13:00	1.91	0.23
6:42	1.62	1.10	9:54	2.92	1.41	13:06	1.88	0.23
6:48	1.69	1.17	10:00	2.93	1.36	13:12	1.85	0.22
6:54	1.76	1.23	10:06	2.92	1.31	13:18	1.82	0.22
7:00	1.83	1.28	10:12	2.88	1.25	13:24	1.78	0.23
7:06	1.90	1.33	10:18	2.87	1.20	13:30	1.75	0.24
7:12	1.96	1.38	10:24	2.85	1.15	13:36	1.70	0.25
7:18	2.02	1.44	10:30	2.82	1.09	13:42	1.67	0.27
7:24	2.08	1.49	10:36	2.81	1.08	13:48	1.64	0.29
7:30	2.15	1.53	10:42	2.81	0.95	13:54	1.63	0.32
7:36	2.20	1.57	10:48	2.79	0.90	14:00	1.63	0.35
7:42	2.26	1.61	10:54	2.76	0.84	14:06	1.61	0.40
7:48	2.31	1.63	11:00	2.72	0.78	14:12	1.59	0.45
7:54	2.35	1.64	11:06	2.70	0.73	14:18	1.54	0.49
8:00	2.40	1.66	11:12	2.65	0.68	14:24	1.51	0.55
8:06	2.45	1.67	11:18	2.60	0.63	14:30	1.50	0.60
8:12	2.50	1.67	11:24	2.56	0.57	14:36	1.50	0.65
8:18	2.54	1.67	11:30	2.52	0.51	14:42	1.50	0.68
8:24	2.57	1.66	11:36	2.48	0.45	14:48	1.50	0.73
8:30	2.61	1.66	11:42	2.44	0.40	14:54	1.49	0.77
8:36	2.64	1.65	11:48	2.39	0.36	15:00	1.49	0.81
8:42	2.67	1.65	11:54	2.35	0.33	15:06	1.49	0.84
8:48	2.72	1.64	12:00	2.31	0.30	15:12	1.53	0.88
8:54	2.76	1.63	12:06	2.28	0.27	15:18	1.53	0.93
9:00	2.79	1.62	12:12	2.24	0.25	15:24	1.57	0.98
9:06	2.80	1.60	12:18	2.20	0.24	15:30	1.59	1.03

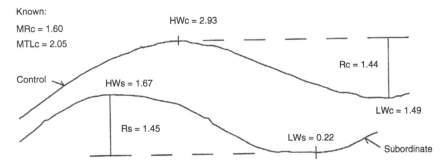

Known:

MRc = 1.60

MTLc = 2.05

HWc = 2.93

Rc = 1.44

Control

HWs = 1.67

LWc = 1.49

Rs = 1.45

LWs = 0.22

Subordinate

Figure 3.3 Typical Standard Method Calculation

station (R_c) being 1.44 (2.93 – 1.49) and the observed range at the subordinate station (R_s) being 1.45 (1.67 – 0.22). With the published values for the 19-year mean range (MR_c) as 1.60 for the control station and the 19-year mean tide level at that station (MTL_c) being 2.05, the equivalent 19-year mean range at the subordinate station (MR_s) may be calculated with Equation 3.1 as follows:

$$\frac{MR_s}{R_s} = \frac{MR_c}{R_c} \text{ which may be restated as } MR_s = MR_c \frac{R_s}{R_c} = (1.60)\frac{1.45}{1.44} = 1.61$$

Next, the average observed low tide (TL_c, TL_s) at both the control and subordinate stations is determined.

$$TL_c = \frac{(HW_c + LW_c)}{2} = \frac{(2.93 + 1.49)}{2} = 2.21$$

$$TL_s = \frac{(HW_s + LW_s)}{2} = \frac{(1.67 + 0.22)}{2} = 0.94$$

The equivalent 19-year mean tide level (MTL_s) at the subordinate station may then be calculated using Equation 3.2.

$$MTL_s = MTL_c - TL_c + TL_s = 2.05 - 2.21 + 0.94 = 0.78$$

Finally, the equivalent values for the 19-year mean high water and mean low water at the subordinate stations may be calculated using Equations 3.3 and 3.4 as follows:

$$MHW_s = MTL_s + \frac{MR_s}{2} = 0.78 + \frac{1.61}{2} = 1.58$$

$$MLW_s = MTL_s - \frac{MR_s}{2} = 0.78 - \frac{1.61}{2} = -0.02$$

3.2.2.3 Example of Tidal Data Calculation (Range Ratio Method)

For an illustration of calculations with the range ratio method, the data from the standard method example (with the assumption that only the upper portion of the subordinate station data was observable) will be used (Figure 3.4).

Using a time interval of four hours, the vertical distance (A_c) below which that interval intersected the curve of the control tide curve was determined to be 0.58 ft and the vertical distance (A_s) below which that interval intersected the curve of the subordinate tide curve was determined to be 0.59 ft (Figure 3.3). The equivalent value for the observed range of the subordinate tidal cycle, had it been observable, may then be calculated using Equation 3.7.

$$R_s = R_c \frac{A_s}{A_c} = 1.44 \frac{0.59}{0.58} = 1.46$$

Once the equivalent value for the observed range at the subordinate station is determined, then the average observed tide at the subordinate station (TL_s) may be calculated using Equation 3.8.

$$TL_s = HW_s - \frac{R_s}{2} = 1.67 - \frac{1.46}{2} = 0.94$$

Then, the average observed tide at the control station (TL_c), the mean range (MR_s), mean tide level (MTL_s), mean high (MHW_s), and mean low water (MLW_s) at the subordinate station calculated by Equations 3.1 through 3.4 as with the standard method.

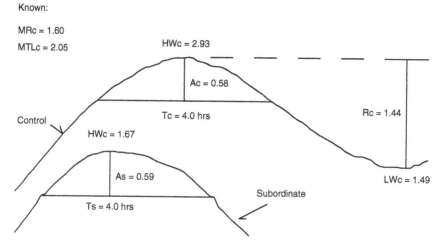

Known:

MRc = 1.60
MTLc = 2.05

HWc = 2.93

Ac = 0.58

Control

Tc = 4.0 hrs

Rc = 1.44

HWc = 1.67

As = 0.59

Subordinate

Ts = 4.0 hrs

LWc = 1.49

Figure 3.4 Typical Amplitude Ratio Method Calculation

$$TL_c = \frac{\left(HW_c + LW_c\right)}{2} = \frac{\left(2.93 + 1.49\right)}{2} = 2.21$$

$$MR_s = MR_c \frac{R_s}{R_c} = \left(1.60\right)\frac{1.46}{1.44} = 1.62$$

$$MTL_s = MTL_c - TL_c + TL_s = 2.05 - 2.21 + 0.94 = 0.78$$

$$MHW_s = MTL_s + \frac{MR_s}{2} = 0.78 + \frac{1.62}{2} = 1.59$$

$$MLW_s = MTL_s - \frac{MR_s}{2} = 0.78 - \frac{1.62}{2} = -0.03$$

As may be seen, there is a slight difference (0.01 in the mean range, mean high, and mean low water values) between the standard and range ratio method, probably due to the form the tide curve not being a perfect sine wave curve. Obviously, the standard method is preferable. Nevertheless, when values for a mean high water datum are needed in areas where only the upper portion of most tidal cycles are observable, this method is a viable alternative.

3.3 Tidal Datum Line Location

Tidal boundaries are lines along a shoreline tracing the local elevation associated with a tidal datum. The traditional method that this is accomplished is based on the assumption that the local datum line is a topographic contour. That contour may be located by leveling and mapped by any of various surveying procedures. Care is necessary when accomplishing this to ensure that significant breaks and inflections in the line are not overlooked in this process. This is especially true when the line lies within marsh or mangrove areas. On open coasts, an innovative approach to mapping the contour is to use tripod-based laser scanning.

Another very practical method for areas without significant wave action is to allow the edge of the water itself to define the line. This approach involves observation of the leading edge of the water on an incoming tide. Typically, when using this process, the elevation of the tidal datum is established on a tide staff in the project area. Then, on an incoming tide, the staff is observed. When the elevation of the tidal datum, such as the mean high water, is reached, the leading edge of the water is flagged and then mapped by any of various surveying methods. One version of this approach that is especially efficient for some areas is to use

Figure 3.5 Tide-synchronized Georeferenced Drone Photograph of the Mean High Water Line in Marsh Area Inundated on Both Sides. Note the Encircled Aerial Targets Placed on the Leading Edge of Water from the Outside Bay. The Site Was Located on a Peninsula. Therefore, Tidal Water Was Also Approaching the High Ground from Inside

tide-synchronized low-level aerial photography taken from a drone platform (Figure 3.5). When that approach is used, care must be taken to select an incoming tide when the sun angle allows graphic viewing of the leading edge of the water. With this method, georeferenced photos can be produced using any of various compilation software packages to allow visual selection of points along the visible edge of water. This approach requires the use of georeferenced aerial targets placed strategically across the flight path for control of the photography. In addition, it is recommended that some targets be placed at random points along the leading edge of the water at the time of the photography. This will guide and document the location of the line.

3.4 Extent of Tidal Influence

At times, the location of the extent of tidal influence in an estuarine river, often called the "head of tide," is necessary. In jurisdictions claiming ownership of only tidal waters, that point would be the extent of public ownership. In other jurisdictions, that point would determine the correct boundary to use upriver from that point. Fortunately, many tidal streams are tidally affected all of the way to their headlands, thus eliminating the necessity for this determination.

Locating the extent of tidal influence can be a time-consuming process. The objective is to locate the point where tidal influence ends with a tide that just reaches mean high water, and very few tides reach exactly that point and then recede. The best process for accomplishing this is to make simultaneous observations at random points up the stream. By observing a few cycles and calculating the tidal range at each point, the approximate position of the inland extent of tidal influence may be estimated. Once that point is located, that position may be refined by observing tides at that point as well as points above and below that point. With reiterations of this process, the inland extent of tidal influence at a mean high tide may be found with reasonable precision.

References

Cole, G.M. (1977). Tidal boundary surveying. Proceedings. American Congress on Surveying and Mapping.

Cole, G.M. (1981). Proposed New Method for Determining Tidal Elevations in Inter-Tidal Zones. Technical Papers. American Congress on Surveying and Mapping, 34–43.

Cole, G.M. (1994). Tidal Water Boundaries. *Stetson Law Review* XX (1): Fall.

Cole, G.M. (1997). *Water Boundaries*. New York: Wiley.

Cole, G.M. (2011). Sea level measurements and their applications. *Proceedings, SaGES Conference*, Mayaguez, Puerto RICO.

Duxbury, A.C. and Alison, B. (1989). *An Introduction to the World's Oceans*. Dubuque, IA: William C. Bron.

Maloney, F.E. and Ausness, R.C. (1974). The use and legal significance of the mean high water line in coastal boundary mapping. *The North Carolina Law Review* 185: December, 1974.

Marmer, H.A. (1951). *Tidal Datum Planes, Special Publication No 135*. Washington, D.C.: U.S. Coast and Geodetic Survey.

Sandars, T.C. (1874). *The Institutes of Justinian*. London: Longmans, Green & Co.

Scott, S. (1931). *Las Siete Partidas*. New York: Commerce Clearing House.

Williams, J. (1962). *Oceanography*. Boston: Little, Brown.

Case Law Cites

Attorney General v. Chambers, 43 Eng. Rep 486 (1854).

Borax Consolidated Ltd. V. City of Los Angeles, 296 U.S. 10 (1935).

Tinicum Fishing Co. v. Carter, 61 Pa. 21, 30–31 (1869).

Lutes v. State, 324 SW 2d 167 (1958).

Ruburt Armstrong v. E.L.A.,D.P.R. 588 (1969). *el Tribunal Supremo de Puerto Rico*.

4

Boundaries in Public Trust Nontidal Waters

4.1 River and Lake Boundary Definitions

Although English common law considered inland, nontidal waters to be owned by the owners of the bordering land, the 1876 U.S. Supreme Court case of *Barney v. Keokuk* ruled that public ownership of navigable waters of the United States extends to inland, nontidal waters as well as tidal waters. Yet, even though all of the states are considered to have received ownership of such waters by virtue of statehood, the states have differing laws regarding the ownership of such waters.

Rivers – Regarding nontidal rivers, a number of the U.S. states follow English common law and accordingly consider the submerged lands beneath navigable rivers to be owned by the adjoining riparian owners. In those states, the ownership of nontidal river front property extends to the center of the stream or river. Those states are reported to include Connecticut, Hawaii, Massachusetts, Maryland, Maine, Michigan, Mississippi, New Hampshire, New Jersey, Ohio, South Carolina, and Wisconsin (Coastal States Organization 1997). In the states that follow that course, as is the case with the English common law, the right of public navigation in such rivers is reserved even though the submerged lands beneath those waters are private land. An example of holdings in these states may be seen with typical case law from Mississippi stating as follows (Archer v. Greenville Sand & Gravel Co.):

> (H)e who owns the bank owns to the middle of the river, subject to the easement of navigation.

In other states, the ownership of adjoining riparian land extends to the ordinary low water mark with the submerged lands below that line owned by the public of the state. Those states are reported to include Alabama, Georgia, Louisiana, Minnesota, and Pennsylvania (Coastal States Organization 1997). An example of

Sea Levels and Coastal Boundaries, First Edition. George M. Cole.
© 2024 John Wiley & Sons, Inc. Published 2024 by John Wiley & Sons, Inc.

holdings in these states may be seen with typical case law from Pennsylvania stating as follows (*Freeland v. Pennsylvania Railroad Company*):

> (I)t has been held in many cases that a survey, returned as bounded by a large navigable river, vests in the owner the right of soil to ordinary low watermark of the stream, subject to the public right of passage for navigation, fishing, etc., in the stream, between ordinary high and ordinary low watermark.

In other states, adjoining riparian ownership along navigable rivers extends only to the ordinary high water mark. Those states reportedly include Arkansas, Florida, Idaho, Illinois, North Carolina, Oregon, Texas, and Washington (Coastal State Organization 1997). An example of holdings in such states may be seen for Florida with typical case law stating as follows (*Martin v. Bush*):

> Upon admission of Florida into the Union . . . the state, by virtue of its sovereignty, became the owner of all lands under the navigable waters within the state, including the shores or spaces, if any, between ordinary low-water mark and ordinary high-water mark

Lakes and Ponds – Regarding the submerged land beneath navigable lakes and ponds, a few states such as Michigan and New York consider the submerged lands below such waters to be owned by the bordering upland owner while most states, including Florida, Idaho, Illinois, Louisiana, New Hampshire, Washington, and Wisconsin consider such water to be part of the public trust with the boundary being the ordinary high water mark. Other states including California, New York, Pennsylvania, Wisconsin, and Vermont also consider such waters as sovereign with the boundary being the ordinary low water mark (Coastal State Organization 1997).

4.2 Nontidal Water Boundary Determination

4.2.1 The Ordinary High Water Mark

A number of judicial decisions provide specific guidance for determining the ordinary high water mark as in the following example (*Tilden v. Smith*):

> High-water mark, as a line between a riparian owner and the public, is to be determined by examining the bed and banks, and ascertaining where the presence and action is so common and usual and long continued in all

ordinary years as to mark upon the soil of the bed a character distinct from that of the banks, in respect to vegetation as well as respects the nature of the soil itself. High-water mark means what its language imports – a water mark.

Such guidance is clear that determination of the boundary should involve an inspection of the soil, vegetation, and geomorphology of the shoreline to find where the character of those features changes from terrestrial to aquatic. This seems straight forward. Yet, in practice, location of the ordinary high water mark is rarely that simple. Many water bodies have a wide flood plain with significant changes in the character of the soil, vegetation, and geomorphology, at not only the upper limit of the bed but also at the upper edge of the flood plain which can be miles distant from the bed of the river. Similarly, multiple escarpments as well as changes in the soil and vegetation may be seen at various elevations around rivers and lakes where there has been stands of water for lengthy periods of time. Therefore, hydrological information is also important to assist in the evaluation and weighing of available evidence to locate a line that agrees with the intent of the law. Considering this, evidence that should be evaluated for determining the appropriate location of the ordinary high water mark may be classified in four categories: geomorphology, botany, soils, and hydrology (Cole et al. 2017). A discussion of various evidence within those categories follows.

Geomorphological Evidence – This includes features indicative of the natural limits of water bodies such as natural levees and escarpments. The margin of most rivers has a similar profile, although the width and slope of the flood plain, foreshore between the ordinary high and low water elevations, and bed may vary considerably with different rivers. One of the most striking features seen for rivers is natural flood-formed levees. *Natural levees* are ridges that parallel a river course, created by the deposition of material when the river overspreads its banks during flood events. In many rivers, flood waters are trapped behind levees in wide flood plains that may be several miles in width. The ordinary high water level is usually found on the river side and below the crest of levees. Erosional features, such as escarpments and significant changes in the character of the soil, may often be found on the waterward side of levees and represent the location of the ordinary high water line.

A pertinent discussion on the use of escarpments for the location of the ordinary high water line is found in *the* **Manual of Instruction for the Survey of Public Land** (Bureau of Land Management 1973) as follows:

Mean (ordinary) high water elevation is found at the margin of the area occupied by the water for the greater portion of each average year. At this level a definite escarpment in the soil is generally traceable, at the top of which is the true position of the meander line. A pronounced escarpment, the result of the action of storm and flood waters, is often found above the principal water level, and separated from the latter by the storm or flood beach. Another, less evident, escarpment is often found at the average low water level, especially of lakes, the lower escarpment being separated from the principal escarpment by the normal beach or shore.

A very refined approach to the use of natural levees and escarpments for determining the boundary of rivers has been developed in Texas (Stiles 1952). That approach has been endorsed by both federal and state courts in that state.[1] The Texas process begins with selection of the "lowest qualified bank" in the area of interest which should be an accretion bank (levee) as opposed to an erosional bank (escarpment). The selected bank should have a well-defined top and bottom, a depression or swale on its landward side, and be the lowest of such banks in the area. The following step is to determine the "basic point," which is the elevation half-way between the top and bottom of the selected bank. The next step is to determine the difference of elevation between the basic point and the current surface of the water at the lowest qualified bank. The ordinary high water line is determined by applying the same difference of elevation to the water level up and down the stream. By this method, the boundary at any point is the same difference above or below the water surface at the lowest qualified bank (Shine 1974).

Levees bordering lakes are less common although most lakes do have escarpments created by wave action. Those may represent the correct location of the ordinary high water mark although there are often multiple escarpments. As a result, collaborating evidence should be used to select the one representing the ordinary high water mark.

Botanical Evidence – The objective of the examination for botanical evidence is to locate the waterward limit of terrestrial vegetation (Figure 4.1). This is sometimes difficult due to the presence of both terrestrial and aquatic vegetation near the shore of water bodies. As a result, care must be taken to determine whether shoreline vegetation is aquatic or terrestrial. The *Manual of Instruction for the Survey of the Public Lands of the United States* offers the following instructions for the distinguishing between aquatic and terrestrial vegetation:

> If vegetation type "A" is found along the water's edge – or even in the water – and type "A" is also found growing at sites situated more toward higher, drier ground (upland), then "A" is a terrestrial species. A good rule

---

1 Examples include *Oklahoma v. Texas, Motl v. Boyd* and *Heard v. State.*

Figure 4.1 Geomorphological (Escarpments), Botanical (Tree) and Soil Change, Evidence of the Ordinary High Water Mark Along Typical North Florida River (Photograph Taken at Very Low Water Level).

of thumb is to determine if the plant is part of a self-reproducing stand of woody vegetation and not a seasonal plant that can sprout and mature in a few months when the water is unseasonably low. Trees, shrubs, and other woody-stemmed plants are generally terrestrial.

Hydrological Evidence – Courts have traditionally recognized the use of botanical, geomorphological, and soils evidence but frowned on the use of mathematical averaging of water levels for determining boundaries of nontidal waters. Typical of this is the U.S. Court of Claims decision in *Kelly's Creek and N.W.R. Co. v. United States:*

> The high water mark is not to be determined by arithmetical calculation; it is a physical fact to be determined by inspection of the river bank.

Yet, there have been cases brought before the courts where evidence based on geomorphological and botanical evidence for opposing sides in the dispute has placed the location of the ordinary high water mark several miles apart. In such situations, hydrological evidence can help determine which line is more reasonable and which agrees with the intent of the law. Generally, such disputes have been regarding whether the ordinary high water line is located at the bank of the water body or at the upper limits of the flood plain. This is the case due to changes in the geomorphology and plant community occurring at both places. An examination of prevailing case law clearly resolves this issue with opinions such as the following:

> It neither takes in overflowed land beyond the bank, nor includes swamps or low grounds liable to be overflowed but reclaimable for meadows or agriculture... (Howard v. Ingersoll; Borough of Ford City v United States)
>
> ... nor the line reached by the water at flood stages (State ex. rel. O'Conner v. Sorenson)
>
> It is the land upon which the waters have visibly asserted their dominion, and does not extend to or include that upon which grasses, shrubs and trees grow, though covered by the great annual rises. (Harrison v. Fite)
>
> The high water mark on fresh water rivers is not the highest point to which the stream rises in times of freshets... (Tilden v. Smith; Dow v. Electric Company)

Further, in many of the public land states, the U.S. Public Land Survey classified many of the flood plains as swamp and overflowed lands with a clear distinction between such lands and submerged lands as follows:

> Swamp lands require drainage to fit them for cultivation. Overflowed lands are those which are subject to such periodical or frequent overflows as to require levees or embankments to keep out the water, and render them suitable for cultivation. (San Francisco Savings Union v. Irwin)
>
> *Overflowed lands include essentially the lower level within a stream flood plain as distinguished from the higher levels...*
> *(Bureau of Land Management 1973).*

As a result, swamp and overflowed lands were typically conveyed to states and in turn subjected to sale to private owners. They are not part of the public domain. Yet, despite this guidance, numerous disputes over flood plains have occurred. In such situations, hydrological evidence can demonstrate which line is more reasonable and which agrees with the intent of the law.

The use of hydrological data is specifically implied in some jurisdictions such as the state of Louisiana where Civil Code Articles 450 and 456 define the boundary between public trust ownership and riparian upland in navigable rivers as the "*ordinary low stage of water*" with the banks subject to private ownership although also subject to a public easement. Further, some case law hints at possible approaches using hydrological evidence with language such as the following:

> This word (ordinary high water)... does not mean the abnormally low level of a lake during one of a series of excessively dry years, or the abnormally high level of a lake during an exceptional wet year or series of wet years, but the average or mean level obtaining (sic) under fairly normal or average weather conditions, allowing the proper range between high and low water mark in average years. (Tilden v. Smith)

Further, there have been some decisions that have encouraged the use of hydrological data due to the need for the precision, repeatability, and lack of ambiguity which results from a mathematical solution. Typical of this are two federal cases in Florida, *U.S. v. Parker* and *U.S. v. Joder Cameron*. In the Cameron case, the court found as follows:

> There is no logical reason why a fourth approach to determining the line or ordinary high water may not consist of comparing reliable water stage and elevation data. Indeed, for a body of water whose levels fluctuate considerably with changes in climate, accurate water stage and elevation data may provide the most suitable method for determining the ordinary high water mark.

Therefore, water level records, where they exist, are potentially as valuable for determining nontidal water boundaries as they are with tidal water boundaries. Yet, resolution of the best method for interpretation of such data is still an unsettled question. General instruction, such as the following from the *Manual of Surveying Instructions* (Bureau of Land Management 2009), gives only very broad instructions on this topic.

> Practically all inland bodies of water pass through an annual cycle of changes and multiyear cycles of drought and wet years. The OHWM is found between these extremes.

Several approaches have been suggested over the years for the use of water levels to support or at least resolve ambiguity in physical evidence. One commonly suggested approach has been the use of the level representing a certain percentage of inundation. That method has not met judicial acceptance due to the arbitrariness of selecting a specific percentage of inundation. Furthermore, evaluation of that method has shown that, depending on the water body, a wide range of percentage values may be found for the ordinary high water line as indicated by more traditional physical indicators.

Nevertheless, there is one rather simple approach to the use of hydro-logical data that would seem to be ideal for use as collaborative evidence to resolve ambiguities between competing physical evidence and as a guide to know where to search for physical evidence. That approach may also serve as the best evidence for the location of the boundary where physical evidence has been obliterated. Such an approach can also be help-ful for the location of the ordinary <u>low</u> water mark where physical evi-dence may be hidden or obliterated by higher water levels.

This process does not involve statistical averaging. Rather, it is method for deter-mining the elevations at which the presence of the water is most common to com-ply with the frequently cited judicial definition for a boundary line "...*where the presence and action of waters are so common and usual and so long continued in all ordinary years...*"

With this approach, available daily water level readings are sorted by elevation. Then, the number of observations occurring in each range of elevations is tabulated. Those data are then plotted in a histogram. That process graphically reveals the range of elevations where the presence of the water has been for long periods of time. For most water bodies, this process reveals two "nodes" where the presence of the water has been most common, one representing a probable elevation for the ordinary high water and one for a probable location for the ordinary low water.

The following example will illustrate an application of the proposed method using water level records for a typical lake. Water level records from a gauge on the lake have recorded daily water level records for over 15 years (Figure 4.2).

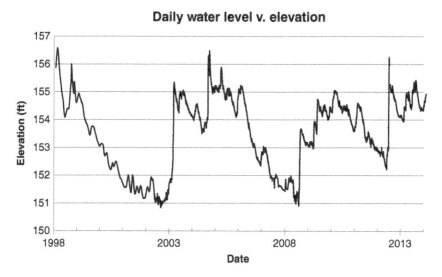

Figure 4.2 Daily Water Level Records.

Table 4.1 Tabulation of Water Levels in Brackets.

Bracket (ft)	No. of days
150–151	42
151–152	**765**
152–153	499
153–154	861
154–155	**1793**
155–156	550
156–157	29

Using brackets of 1 ft of elevation, the daily elevations observations were sorted by elevation, and a count made of the number of days where the water level was in each foot of elevation during the period of record (Table 4.1). Those counts were then used to create a histogram (Figure 4.3). Unlike many histograms which use columns, a linear graph is used for better resolution for this application.

As may be seen, the histogram graphically depicts two "nodes" or elevation brackets where the "presence and action of the water" is most prevalent. The higher of these, at an elevation between 154 and 155 ft, would suggest the elevation at which to search for evidence of the ordinary high water mark. The lower, at an elevation between 151 and 152 ft, would suggest the elevation at which to search for evidence of the ordinary low water mark.

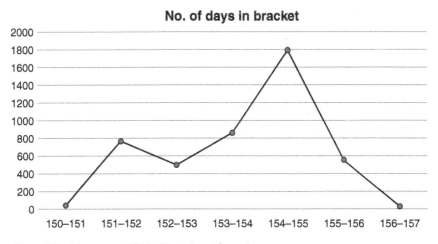

Figure 4.3 Histogram of Daily Water Level Records.

Tests of this approach suggest that the above example, is somewhat of an anomaly in the resulting histograms. In other tests of this approach, the node for the ordinary high water is far less prominent than that for the ordinary low water. As an example, an analysis was made using 24 years of hydrographic data from a USGS gauge located in the Withlacoochee River near Quitman, Georgia (Figure 4.4). An analysis was made of the number of days that the water level was in each of the 1-ft elevation brackets (Table 4.2) and a histogram created from that tabulation (Figure 4.5).

As may be seen, the analysis indicated a prominent low water node in the 86–87 ft bracket. In addition, the analysis indicated a much smaller node at the 89–90 ft elevation associated with high water.

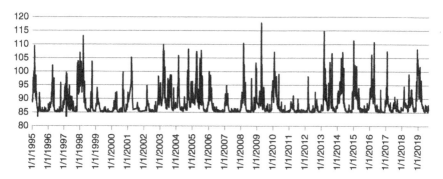

Figure 4.4 Plot of Available Data.

Table 4.2 Tabulation of Days in Each Elevation Bracket.

Bracket	Days	Bracket	Days	Bracket	Days
85–86	463	97–98	88	109–110	7
86–87	**2950**	98–99	83	**110–111**	**6**
87–88	817	99–100	80	111–112	2
88–89	622	100–101	85	112–113	2
89–90	617	101–102	42	113–114	20
90–91	315	102–103	47	114–115	4
91–92	246	103–104	43	115–116	1
92–93	214	104–105	30	116–117	1
93–94	199	105–106	23	117–118	1
94–95	127	106–107	21	118–119	2

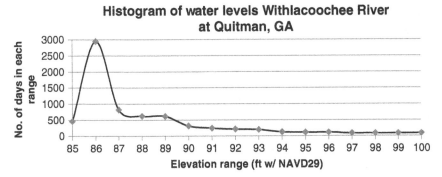

Figure 4.5 Histogram. *Source:* www.tidesandcurrents.noaa.gov.

Field examination of the river near the gauge revealed a prominent low water escarpment with a top at approximately 86 ft in elevation. In addition, a prominent escarpment with a base elevation of about 88.5 and a top elevation of 90 ft was observed on the waterward side of a prominent levee was along with various other indicators of the ordinary high water mark including the landward extent of mature trees and change in the character of the soil. In addition, a sharp low water escarpment with a base of 85.2 and a top of 86.2 was observed.

Field tests of the approach on a number of water bodies have been positive in that there is agreement between the nodes revealed in the analyses and physical evidence, although care must be taken in the interpretation of the results. As an example of the agreement, the process graphically predicted the elevation of the ordinary high water mark at the pronounced escarpment located at the waterward extent of upland vegetation as illustrated in Figure 4.1 earlier in this section.

4.2.2 The Ordinary Low Water Mark

Although several of the states recognize the ordinary low water mark as a boundary for nontidal sovereign water, legal guidance as to how to identify such a boundary is scarce. Considering the four categories of evidence for the ordinary high water mark (geomorphology, botany, soils, and hydrology), only geomorphology and hydrology is practical for the ordinary low water mark. Therefore, the best traditional evidence usually consists of escarpments or other significant changes in slope representing the point where the sloping foreshore changes to the level of the bed of the water body. Unfortunately, such evidence of the ordinary low water line is underwater much of the time. The previously described process using water level records has proven to be an excellent tool for providing evidence of the location of the ordinary low water mark since it can identify the probable location of that mark, even if underwater at the time of the investigation.

4.3 Use of Government Land Office Meander Lines as Boundaries

In some cases dealing with boundaries of rivers and lakes in the public land states, it has been suggested that meander lines established in the Government Land Office survey be used as boundaries. The following section[2] analyses that approach:

One of the more important charges to government surveyors performing the original surveys of public lands of the United States was to separate the upland, subject to disposal under the public land laws, from the navigable waters, which were to be preserved as public highways. Even a century or two later, the decisions of these surveyors take on a great deal of significance in disputes regarding lands lying under and bordering on water bodies. The separation process involved meandering, or running a series of measured courses and distances, around the perimeter of the water body. The alternate course of action, for non-navigable water bodies, was to continue section lines across the water bodies. For such non-navigable water bodies, the width was usually noted in the field notes and the approximate limits of the water sketched on the plat, as opposed to an actual survey or meander of the shoreline, as required for navigable water bodies.

4.3.1 To Meander or Not to Meander

Pertinent Instructions – The meandering of navigable water bodies has been an integral part of the survey of the public lands since its inception (Bouman 1977). As an illustration of this, the first survey conducted under the rectangular survey system (Township 1, Range 1, Seven Ranges, Ohio by Deputy Surveyor Absolom Martin) included meandering of the Ohio River. However, the early laws detailing procedures for public land surveys did not address meandering as such. They did, however, mention the reservation of navigable rivers. This is illustrated by the following:

> That all navigable rivers, within the territory to be disposed of by virtue of this act, shall be deemed to be, and remain public highways. (Section 9, Act of May 18, 1796).

The earliest detailed instructions for the public land survey were those issued by Surveyor General (Northwest of the Ohio) Edward Tiffin in 1815. These instructions

---

2 This section was originally written as an article in *Surveying and Land Information Systems*, Vol 50, No. 3, September 1990, American Congress on Surveying and Mapping (Cole 1990) and is reprinted with permission.

did provide some direction as to which waters to meander, at least for rivers, but did not address lakes. The instructions read as follows:

> The courses of all navigable rivers, which may bound or pass through your district, must be accurately surveyed...

The first mention of lakes appears to be in the instructions for the survey of lands in the state of Mississippi, issued in 1832 by Gideon Fitz, surveyor general of the lands south of Tennessee. In 1842, instructions to deputy surveyors in Florida also mentioned lakes:

> You will accurately meander, by course and distance, all navigable rivers which may bound or pass through your district; all navigable bayous flowing from or into such rivers; all takes and deep ponds of sufficient magnitude ...

The first general nationwide manual of instructions, issued in 1855, provided even more specific language regarding rivers, and included a requirement to meander all lakes of 25 acres or larger. A published supplement to the 1855 manual, written in 1864, modified the manual to increase the lower limit for lakes to 40 acres. In addition, a new class of non-navigable river was mentioned. The instructions related to those two items read:

> Rivers not embraced in the category denominated 'navigable' under the statute, but which are well-defined natural arteries of internal communication and have a uniform width, will be meandered on one bank ...
> Lakes embracing an area of less than forty acres will not be meandered. Long, narrow or irregular lakes of larger extent, but which embrace less than one-half of the smallest legal subdivision, will not be meandered. Shallow lakes or bayous, likely in time to dry up or be greatly reduced by evaporation, drainage, or other cause, will not be meandered, however, extensive they may be.

The requirement for meandering of one bank for non-navigable "arteries of internal communication" was also included in the 1881 manual and then dropped in later manuals. The 1881 manual also did away with the requirements to meander all lakes of 40 acres and above, and instead required that "*all lakes, bayous, and deep ponds which may serve as public highways of commerce*" be meandered. Those instructions also required that "lakes, bayous, and ponds lying entirely within a section are not to be meandered." The 1890 manual of instructions reinstated the requirement that all lakes over 25 acres were to be meandered. This restriction was also included in the 1894, 1902, 1919, and 1930 manuals. All size restrictions were deleted in the last two (1947 and 1973) manuals.

The 1890 manual also instituted the so called three-chain rule for streams. This instruction required the meandering of streams wider than three chains (198 ft) wide, regardless of their navigability. This rule was also included in the 1894, 1919, and 1930 manuals, but was deleted in later manuals.

4.3.2 Legal Significance of Meandering

As may be seen from the previous example, federal surveyors made an effort to identify all navigable waters in public lands. Since these surveys were often performed relatively near the time of statehood, they reflect navigability determinations of the water bodies in the condition they were in at that time. This would, therefore, be the deciding factor in determining which water bodies were received by the states by virtue of statehood. Because of this, these surveys often represent the most reliable and often the only inventory of navigable waters in most public land states. Therefore, the courts have relied upon the public land surveys for a presumption of navigability or non-navigability. An example of this is a recent Florida Supreme Court case (*Odom v. Deltona*):

> In Florida, meandering is evidence of navigability which creates a rebuttable presumption thereof. The logical converse of this proposition, noted by the lower court, is that non-meandered lakes and ponds are rebuttably presumed non-navigable.

Obviously, the early surveyors were not infallible – any more than surveyors are today. As a result, their decisions regarding the navigability or non-navigability of a water body were not always correct. As indicated in the previous opinion, such evidence creates only a presumption, which may be rebutted by evidence to the contrary. Nevertheless, these surveys are obviously the best evidence of the navigability of a water body at statehood in most cases, and they have, therefore, received judicial recognition as such.

It is noted, however, that when considering the evidence of meandering in government surveys, one must be aware of the prevailing instructions controlling that survey. For example, the occasional requirement that lakes down to a size of 25 acres be meandered, regardless of their navigability, would indicate that a meandered lake would not necessarily be considered as navigable.

4.3.3 Location of Meander Lines

4.3.3.1 Pertinent Instructions

Although, as has been demonstrated in previous sections, meandering was mentioned in some of the earliest laws and instructions, directions as to placement of the meander line were not mentioned in the Manual of 1881. Salient portions of that manual follow:

> Meander lines should not be established at segregation line between dry and swamp or flowed land, but at the ordinary low-water of the actual margin of the rivers or lake border ...
>
> In the survey of lands bordering on tide water, 'meander corners' are to be established at points where surveyed lines intersect high-water mark, and meanders are to follow the high-water line.

These instructions remained the same in the 1890 manual. In the 1894 manual, the instructions remained the same, except that the term "ordinary low water mark" (in the first paragraph above) changed to "ordinary high water mark." The 1919 manual reflected considerably more detail in the meandering section and contained instructions that have remained essentially the same in subsequent manuals. Salient portions of those instructions follow:

> All navigable bodies of waters and other important rivers and lakes (as hereinafter described) are to be segregated from the public lands at mean high water elevation.
>
> Mean high-water mark has been defined in a State decision (47 Iowa, 370) in substance as follows: High-water mark in the Mississippi River is to be determined from the river bed; and that only is riverbed which the river occupies long enough to wrest it from vegetation. In another case (14 Penn. St., 59) a bank is defined as the continuous margin where vegetation ceases, and the shore is the sandy space between it and low water mark.
>
> Meander lines will not be established at the segregation line between upland and swamp or overflowed land, but at the ordinary high water mark of the actual margin of the river or lake on which such swamp or overflowed lands border.
>
> Practically all inland bodies of water pass through an annual cycle of changes from mean low-water to flood stages, between the extremes of which will be found mean high-water.
>
> The surveyor will find the most reliable evidence of mean high-water elevation in the evidence made by the water's action at its various stages, which will generally be found well marked in the soil, and in timbered locations a very certain indication of the locus of the various important water levels will be found in the belting of the native forest species.

Mean high-water elevation will be found at the margin of the area occupied by the water for the greater portion of each average year; at this level a definite escarpment in the soil will generally be traceable, at the top of which is the true position for the surveyor to run the meander line.

4.3.3.2 Legal Significance of Location

As may be seen from the previous example, the earliest instructions regarding the location of meander lines (1881) indicate that these lines were to be established at what we today consider to be the ordinary high water mark. It may be assumed that those instructions represented the codification of previous practices. Therefore, this probably had been the practice since the inception of the public land surveys.

The detailed instructions in the later manuals are for a line that is identical to ordinary high water lines located by today's practicing surveyors in nontidal waters. In tidal waters, the instructions obviously will not yield as precise and repeatable a line as a more sophisticated mean high-water line determined by modern techniques. Nevertheless, in most cases, the line found by such instructions would be a good approximation of the mean high water line. Therefore, it appears that the meander line was at least intended to be, and should be in most cases, a reasonable approximation of the ordinary high water line or mean high water line at the time of the survey. Water boundaries are often ambulatory, so this obviously would affect the use of meander lines as current boundaries. This is reflected in a number of court opinions typified by the following:

> It is an established and accepted principle, subject only to the exception hereinafter noted, that the meander line of an official government survey does not constitute a boundary, rather the body of water whose shoreline is meandered is the true boundary . . . (Connerly v. Perdido Key, Inc.).

Meander lines are significant, however, in that they may be used to represent the approximate location of the ordinary high water line at a certain point in history. This is often a factor in determining the original location of altered shorelines. In addition, there may be other instances where the current water boundary cannot be located and the meander line may be accepted as the boundary by default, or where the discrepancy between the meander line and the shoreline is large enough to indicate intentional omission of certain lands or fraud. These instructions are illustrated by the following exceptions from court opinions:

> Under the circumstances of this case [Testimony had indicated that the true mean high-water line could not be located with any certainty], we hold that the meander line constituted the boundary line between

the swamp and overflowed lands and the sovereignty lands ... (Trustees v. Wetstone).

However, a meander line may constitute a boundary where so intended or where the discrepancies between the meander line and the ordinary high-water line leave an excess of unsurveyed land so great as to clearly and palpably indicate fraud or mistake (Lopez v. Smith).

Therefore, although certain limitations must be recognized, the location of a meander line may have significant legal significance regarding boundary location.

4.4 Case Studies for Nontidal Boundaries (Cole 1997)

To illustrate typical applications of the principles for nontidal water boundaries, two examples will be provided. These will include two prominent judicial rulings dealing with such boundaries, one a federal case and one a state case.

Case Study 1   Howard v. Ingersoll

A study of this case is instructive in that it is the source of the modern definition of the ordinary high water line. Further, it represents the initial attempt of the U.S. Supreme Court in defining the line. The case can be confusing because it contains three, concurring but sometimes confusing, opinions. Nevertheless, the case does represent a definite step in the evolution of the modern definition of the ordinary high water line and is therefore important for a clear understanding of that definition when coupled with later, more definitive, opinions. At issue in this case was the meaning of a deed call for the boundary between the states of Alabama and Georgia which was described as running up the western bank of the Chattahoochee River. The main opinion of the Court in this case defined the line as follows:

> When banks of rivers were spoken of, those boundaries were meant which contain their waters at their highest flow, and in that condition, they make what is called the bed of the river ... it neither takes in overflowed lands beyond the bank, nor includes swamps or low grounds liable to be overflowed, but reclaimable for meadows or agriculture.

(Continued)

Case Study 1   (Continued)

> ...The call is for the bank, the fast land which confines the water of the river in its channel or bed in its whole width, that is to be the line. The bank or the slope from the bluff or perpendicular of the bank may not be reached by the water for two thirds of the year, still the water line impressed upon the bank above the slope is the line ...

According to a noted water law authority (Maloney 1978), this explanation *"was confusing and tended towards what with hindsight we would call over-breadth ... the main problem with this version is that it apparently contemplates drawing the line at the 'highest flow' or stage of the river."* It is noted that the point of highest flow appears to conflict with the exclusion of overflowed lands from the bed of the river, which is also called for in this passage.

A concurring opinion, written by Justice Nelson and concurred to by Justice Grier, placed the line at the ordinary level as follows:

> In the ordinary state of the river ... the water covers this flat about halfway to the high bluff, extending to the base of a bank or ridge of sand or gravel: and in freshets, the water covers the flats reaching to the bluff. (The line) does not necessarily, nor as I think, reasonably, call for a line along the bluff or high bank... The bank enclosing the flow of water, when at its ordinary and usual stage, is equally within the description; and the limit within this bank, on each side, is more emphatically the bed of the river, than that embraced within the more elevated banks when the river at flood... In our judgement, the true boundary line ... is the line marked by the permanent bed of the river by the flow of the water at its usual and accustomed stage ...

A second concurring opinion by Justice Curtis added greater definition to the boundary in this case by defining it "by reference to several ascertainable physical characteristics of the bank" (Maloney 1978) as follows:

> (The) line is to be found by examining the bed and banks, and ascertaining where the presence and action of water are so common and usual and long continued in all ordinary years, as to mark upon the soil of the bed a character distinct from that of the banks, in respect to vegetation, as well as in respect to the nature of the soil itself. Whether this line ... will be found above or below, or at a middle stage of water, must depend upon the character of the stream ... But in all cases the bed of the river is a natural object ... the banks being fast land, on

which vegetation, appropriate to such land in the particular locality, grows wherever the bank is not too steep to permit such growth, and the bed being soil of a different character and having no vegetation, or only such as exists when commonly submerged in water.

Later cases, as previously suggested, appear to have followed the two concurring opinions as opposed to the primary opinion in this case.

Case Study 2   Tilden v. Smith

The case of Tilden v. Smith represents a typical state court decision dealing with the ordinary high water line. This case was decided by the Supreme Court of Florida in 1927. It involved a dispute over the rights of littoral owners of Lake Johns in Orange County, Florida. The actual question before the court was whether one riparian owner had the right to lower the level of the lake, when it was in an overflowed stage, by use of a drainage well. Although the location of the ordinary high water line was not the primary question addressed by this case, a review of the opinion, coupled with knowledge of the lake involved, allows insight into what the court considered to be the ordinary high water line and especially the treatment of the flood plain.

Lake Johns, unlike many Florida lakes, is not fed by underground springs, but by rainfall and runoff. As a result, there is considerable variation in water level from year to year due to meteorological cycles. In this case, the court took note of this large variation but stated that "... *nevertheless the character of the vegetation and trees around the lake gave some evidence of an average or ordinary high water mark* ..." At the time of the litigation, the lake was at an abnormally high stage. A golf course of the West Orange County Club, which was located on the flood plain of the lake, was submerged and nearby cottages made inhabitable by the high water level.

A deep well drilled for drainage purposes was, according to evidence introduced, situated "at a point higher than the level of the lake within its natural boundaries." The disputed well, in addition to a second well drilled at approximately the same time, has been recovered and its elevation to a nearby vertical control point. That process revealed elevations of the casings of the two wells of 92.6 and 92.8 ft above the National Geodetic Vertical Datum (Gentry 1989).

An examination of the shore of the lake provides both botanical and geomorphological evidence of the ordinary high water line. The geomorphological evidence includes pronounced escarpments at elevations of

(Continued)

Case Study 2   (Continued)

approximately 98.5 and 90.0 ft, NGVD. In addition, a low water escarpment was found at an elevation of 88 ft (Gentry 1989). That profile matches the classic profile described in the previously provided quote from the *Manual of Instruction for the Survey of Public Lands* as follows: *Practically all inland bodies of water pass through an annual cycle of changes and multiyear cycles of drought and wet years. The OHWM is found between these extremes.* Based on such evidence, the ordinary high water line would be along the middle escarpment, which is slightly below the elevation of the wells described in the court opinion.

That location is supported by vegetative evidence. The flood plain of Lake Johns is relatively devoid of upland trees, apparently due to the nature of the water level fluctuations of the lake. The apparent ordinary high water line, based on vegetation, is at the waterward edge of scattered shrublike vegetation. There is also a distinct change in the grasses at that point with more water-tolerant species emerging. Leveling indicates that this occurs at an elevation of 90 ft (Gentry 1989), which is consistent with the geomorphological evidence as well as the court's holding of the location of the line.

References

Bouman, L. (1977). The meandering process in the survey of the public lands of the United States. In: *Proceedings of the Water Boundary Workshop*. California Land Surveyors Association.

Bureau of Land Management (1973). *Manual of Instructions for the Survey of the Public Lands of the United States*. Washington, D.C.: U.S. Government Printing Office.

Bureau of Land Management (2009). *Manual of Instructions for the Survey of the Public Lands of the United States*. Washington, D.C.: U.S. Government Printing Office.

Coastal States Organization (1997). *Putting the Public Trust Doctrine to Work*. Washington, D.C: Coastal States Organization.

Cole, G., Hale, J., Ward, D., and Joanas, Z. (2017). *Use of bathymetric LIDAR for paleo landscape description*. In: *Lost and Future Worlds Conference*. Buckinghamshire, UK: Royal Society.

Gentry, D. (1989). *Personal Communication*. Orlando, FL: Jones Wood & Gentry.

Shine, D. (1974). Personal Communication, Silsbee, TX.

Stiles, A. (1952). The gradient boundary. *Texas Law Review* 30 (3): University of Texas, Austin, TX.

Case Law Cites

Archer v. Greenville Sand and Gravel, 233 U.S. 60 (1914).

Borough of Ford City v. United States, 345 F. 2d 645 (1965).

Connerly v. Perdido Key, Fla., 270 So. 2d 390 (1972).

Dow v. Electric Company, N.H., 45 A 350 (1899).

Harrison v. Fite, 148 F 781 (1906).

Heard v. State, Tex., 204 S.W. 2d. 795 (1957).

Howard v. Ingersoil, 54 U.S. 381, 427 (1851).

Kelly's Creek and N.W.R. Co. v. United States, 100 Ct. CI 396 (1943).

Lopez v. Smith, 109 So. 2d 176, Fla. 2d. DCA (1959).

Martin v. Bush, Fla.,112 So. 274 (1947).

Motl v. Boyd, Tex., 286 W.W. 458 (1926).

Odom v. Deltona, Fla.,341 So. Ed. 977 (1976).

San Francisco Savings Union v. Irwin, 28 F 708 (1886).

State ex rel O'Connor v. Sorrenson, Okla. 198 P 2d. 402.

Tilden v. Smith, Fla., 113 So. 708 (1927).

Trustees v. Wetstone, Fla., 222 So. 2d. 10 (1969).

U.S. v. Joder Cameron, 466 Fed. Supp. 1099 (1979).

U.S. v. Parker, No. 75-34, N.D. Fla (1976).

5

U.S. National and State Maritime Boundaries

5.1 National Maritime Boundaries

Prior to the eighteenth century, most international shipping recognized informal territorial sea boundaries bordering coastal nations. Generally, the width of such areas was considered to be about 3 mi based on the informal "cannon shot rule." In the latter part of that century, France proposed a formal standard width of 3 nmi for purposes such as the right of passage for other nations and rights to fishing and natural resources. Other maritime nations, including the newly formed United States, agreed to that standard in a 1793 treaty. Following that, most nations recognized the 3 nmi limit resulting from that treaty.

Yet, as changes in technology allowed great use of the sea, nations such as Denmark which were highly dependent on fishing began to claim a 50 nmi fisheries limit. Nevertheless, the United States made no attempt to extend its jurisdiction beyond the 3 nmi limit until the 1930s. At that time, the U.S. jurisdiction was extended to 12 nmi for law enforcement purposes, primarily to stop offshore rum runners during the prohibition era. That jurisdiction extension was of an extraterritorial nature and did not extend territorial sea ownership.

After the discovery of oil offshore of California in 1945, the jurisdiction of the United States was extended even farther seaward with a proclamation by President Truman stating that "...the Government of the United States regards the natural resources of the subsoil and seabed of the continental shelf[1] beneath the high seas but contiguous to the coasts of the United States as appertaining to the United States, subject to its jurisdiction and control" (*Truman 1945; 43 USC Sec. 1332(l)*). Again, that jurisdictional extension was of an extra-territorial nature.

---

1 The continental shelf is the gently sloping plain of land along the coasts of most continents and islands. It varies greatly in width, from a few miles to hundreds of miles. It is considered to end where the continental slope begins to drop more steeply to the ocean floor.

Sea Levels and Coastal Boundaries, First Edition. George M. Cole.
© 2024 John Wiley & Sons, Inc. Published 2024 by John Wiley & Sons, Inc.

Following the U.S. extension, claims for natural resources and fishing limits began to vary widely worldwide. As one example, Iceland, to protect its vital fishing industry, claimed a 50 nmi fishing jurisdiction. This was hotly disputed by the British fishing industry which had traditionally fished within those limits. This resulted in a bitter and long continued dispute between the two nations called the "cod wars" which included at least one fatality. While on assignment in that nation in the 1960s in connection with establishing a worldwide geodesy network, the author saw graphic evidence of that struggle in an Icelandic government's aircraft hangar in the form of a British whaleboat mounted as a trophy of the cod wars.

The issue of a standard territorial sea has been frequently debated at international conferences. At a 1930 Hague Conference of the International Law Commission and at 1958 and 1973 United Nations Conferences on the Law of the Sea, attempts were made to adopt a standard width zone without success.

In 1977, as a result of widely varying claims by other nations and the failure of the 1973 United Nations Law of the Sea Conference to develop international standards, the United States, formerly claimed exclusive fishing management jurisdiction to a 200 mi limit, regardless of whether this line went beyond the continental shelf (*16 USC Sec. 181 1*). Nevertheless, the United States continued to maintain a 3-mi territorial sea claim although other nations claimed greater widths for their territorial seas.

In 1982, a United Nations conference succeeded in establishing a standard width for territorial seas. That conference adopted an international law of the sea treaty providing for coastal nations having a territorial sea running out to 12 mi from shore with exclusive economic zones with a width of 200 mi or to a maximum of 350 mi if the country's continental shelf extended out that distance. The limit of the continental shelf was defined as out to "*a depth of 200 m or, beyond that limit, to where the depth of the superjacent water admits of the exploitation of the natural resources of the said area...*" (Figure 5.1).

The coastline used as a baseline from which to measure for establishing national boundaries is the low water line depicted on charts of the nation (which in the United States is currently the mean lower low water line). To address situations where there are indentations in the coastline or when islands or other unusual coastline configurations are encountered, the United Nations Law of the Sea Conferences have developed rigid guidelines for baselines. Those guidelines have also been adopted by case law for state/federal boundaries as outlined later.

Article 76 – Definition of the continental shelf

(1) The continental shelf of a coastal sate comprises the seabed and subsoil of the submarine areas that extend beyond its territorial sea throughout the natural prolongation of its land territory to the outer edge of the continental margin, or to a distance of 200 nmi from the baselines from which the breadth of the territorial sea is measured where the outer edge of the continental margin does not extend up to that distance.

(2) The continental shelf of a coastal state shall not extend beyond the limits provided for in paragraphs 4–6.

(3) The continental margin comprises the submerged prolongation of the land mass of the coastal state and consists of the seabed and subsoil of the shelf, the slope, and the rise. It does not include the deep ocean floor with its oceanic ridges or the subsoil thereof.

(4) (a) For the purposes of this convention, the coastal state shall establish the outer edge of the continental margin wherever the margin extends beyond 200 nmi from the baselines from which the breadth of the territorial sea is measured, by either:

   (i) A line delineated in accordance with paragraph 7 by reference to the outermost fixed points at each of which the thickness of sedimentary rocks is at least 1 per cent of the shortest distance from such point to the foot of the continental slope; or

   (ii) (b) In the absence of evidence to the contrary, the foot of the continental slope shall be determined as the point of maximum change in the gradient at its base.

(5) The fixed points comprising the line of the outer limits of the continental shelf on the seabed, drawn in accordance with paragraph 4 (a)(i) and (ii), either shall not exceed 350 nmi from the baselines from which the breadth of the territorial sea is measured or shall not exceed 100 nmi from the 2500 m isobath, which is a line connecting the depth of 2500 m.

(6) Notwithstanding the provisions of paragraph 5, on submarine ridges, the outer limit of the continental shelf shall not exceed 350 nmi from the baselines from which the breadth of the territorial sea is measured. This paragraph does not apply to submarine elevations that are natural components of the continental margin, such as its plateaux, rises, caps, banks, and spurs.

Figure 5.1 Definition of the Continental Shelf. *Source:* www.tidesandcurrents.noaa.gov.

1) When a river flows directly into the sea, the baseline is in a straight line across the river between headland points on the low tide line on either side of the mouth of the river.
2) The baseline at the mouth of bays, as will be discussed in Chapter 8, runs along the closing line drawn between the headlands of the bay. This is the case for either a juridical or historic bay.
3) Where low tide elevations exist within the territorial sea, when measured from the mainland or an island, then the low water line for those elevations may be used as the baseline. A low tide elevation is defined as *"a naturally formed area of land which is surrounded by and above water at low tide but submerged at high tide" (Article 3, 1958 United Nations Convention on the Law of the Sea).*
4) An island is defined as a *"naturally formed area of land, surrounded by water, which is above water at high tide" (Article 10, 1958 United Nations Convention on the Law of the Sea).*

The United States chose not to become a party to the Law of the Sea treaty, reportedly due to disagreements over how deep-sea mining should be administered and its profits shared. Nevertheless, the United States did adopt an equivalent-width territorial sea in 1988 when President Ronald Reagan extended the U.S. territorial sea *"... to the limits permitted by international law..."* with a presidential proclamation (Reagan 1988) which read as follows:

> The territorial sea of the United States henceforth extends to 12 nmi from the baseline of the United States determined in accordance with international law.

As a result of all of the above, the United States currently claims exclusive ownership of a territorial sea of 12 nmi from a baseline along the mean lower low water line of its coastline; exclusive jurisdiction over the natural resources to the extent of the continental shelf; and exclusive fishing rights out to 200 mi.

5.1.1 Bays

As mentioned in previous sections, the baseline for measuring out to a seaward boundary of the United States and of the individual coastal states is the mean lower low water line or closing lines drawn across the mouths of rivers or bays. As a result, whether a coastal indentation is a bay or merely a part of the open sea is an important question when dealing with federal, state, and international offshore boundaries. If the indentation is a bay and therefore part of the inland waters, then the baseline for delineating the offshore boundary is a straight closing line across the entrance. If the indentation is part of the open sea, the baseline follows the shoreline into the indention which may result in differences in the ownership and control of large areas of offshore submerged land.

The U.N. Convention on the Territorial Sea and the Contiguous Zone defines a bay as a *"well marked indentation where penetration is in such proportion to the width of its mouth as to contain landlocked waters and constitute more than a mere curvature of the coast."* As a result, the navigability or depth of the water in a bay is irrelevant. The treaty further limited the category of bays to indentions where the entrance does not exceed 24 nmi.

As a result, the fact that a water boundary is called a bay does not necessarily make it a juridical bay (a bay in the legal sense). As may be seen from the definition, the extent of the penetration of the waters into the land, in proportion to the width of the entrance would appear to be the major criterion. This was defined in the treaty in objective mathematical terms by the so-called "semi-circle rule" as follows: *"An indentation shall not, however, be regarded as a bay unless its area is as large as, or larger that, that of the semi-circle whose diameter is a line drawn across the mouth of that indentation."*

One class of indentations which is regarded as exceptions to the rule defining juridical bays are historic bays. These are waters where, through long standing assertion of rights and acquiescence by others, a nation or state has established title by prescriptive. The purpose of this exception is to exclude from consideration *"…certain bays whose status has been already settled by history"* (Shalowitz 1962). There are no limitations on the width of the entrance or the relationship of the depth to entrance width for historic bays.

In addition to closing lines across the mouth of bays serving as a baseline for state and national offshore boundaries, the limits of bays sometimes serve for other legal purposes. As an example, some governmental regulations apply within the limits of bays. With the outer boundary being defined by a closing line as previously discussed, the remaining critical boundary is the upper limit of the bay. That limit is typically considered to be a line connecting the headlands of all tributaries to the bay. Therefore, a bay would be considered to include all of the submerged land between the head of the bay defined in that fashion and the closing line across the mouth of the bay.

5.1.2 Entrance Points (Headlands)

As previously discussed, the seaward limit of rivers and bays is a line between headlands. As a result, the selection of acceptable headlands is an important consideration. There is typically no controversy on such selections along coastlines with pronounced physical features. Nevertheless, for some water bodies, especially along low-dynamic waters such as that of the U.S. Gulf of Mexico coastline, there can be a surprising variety of choices. Webster's dictionary defines a headland as *"a point of usually high land jutting out into a body of water."* Yet, in the context used for water boundaries, the horizontal dimension of the headland is the critical factor.

For purposes of maritime boundaries, a headland is defined as "...*the apex of a salient of the coast; the point of maximum extension of a portion of the land into the water; or a point on the shore at which there is appreciable change in direction of the general trend of the coast*" (Shalowitz 1962). The meeting of the forces of the ocean and those of the estuary typically forms a characteristic feature, such as a sand spit or cusp, which is the headland sought. The critical feature is the outermost extension of the headland.

If the headland does not have a distinct point and is more rounded, a point should be selected by bisecting the angle formed by the lines following the general trends of the low water line of the open coast and the low water line of the estuary (Figure 5.2). This approach was suggested by the U.S. Supreme Court (*U.S. v. California*) as follows:

> Where there is no pronounced headland, the line should be drawn to a point where the line of mean lower low water on the shore is intersected by the bisector of the angle formed where a line projecting the general trend of the line of mean lower low water along the open coast meets a line projecting the general trend of the line of mean lower low water along the tributary estuary.

An objective method of selecting the correct headland is the use of a "45 degree test" (Hodgson and Alexander 1972). If the angle between the general direction of the shoreline within the bay or river and the proposed closing line across the opening is 45 degrees or greater, then by this test, the correct headland has been

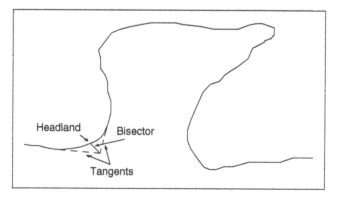

Figure 5.2 Bisector of Angle Approach for Locating Headlands. *Source:* Adapted from Hodgson and Alexander (1972).

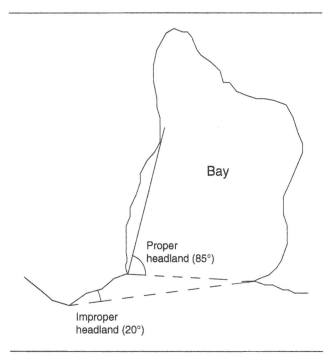

Figure 5.3 45° Test for Locating Appropriate Headlands (Cole 1997).
Source: www.tidesandcurrents.noaa.gov.

selected (Figure 5.3). This test has also been mentioned by the U.S. Supreme Court in dictum (*U.S. v. Maine*).

5.1.3 Equidistant/Median Lines

In addition to defining the seaward boundary of coastal waters, maritime boundaries include lateral dividing lines between adjacent nations or states. If all coastlines were straight and their upland boundaries perpendicular to the coastlines, then equitable lateral boundaries would simply be at right angles to the coast. But since that is rarely the case, lateral boundaries at right angles typically create inequitable division of the territorial sea. The solution agreed upon by the International Law Commission is a median (equidistant) line, *"every point of which is equidistant from the nearest points on the baselines from which the breath of the territorial seas of the two States is measured."*

For typical configurations (Figure 5.4), an equidistant (median) line from the termination of the land boundary between adjacent states to the limits of the territorial sea may be constructed as follows (Shalowitz 1962):

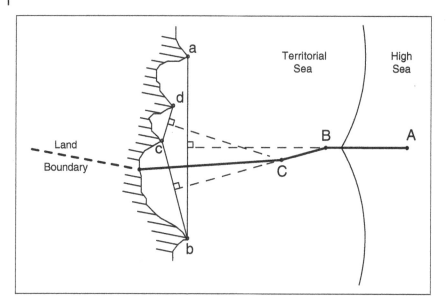

Figure 5.4 Construction of Equidistant Line Between Adjacent Coasts. *Source:* Shalowitz (1962)/NOAA/Public Domain.

(1) Identify salient points (*a* and *b*) on either side of the coastal termination of the upland boundary whose location has been used to measure the breadth of the territorial sea.

(2) Select a point (*A*) which is outside of the territorial sea and equidistant from salient shoreline points, one on either side of the termination of the land boundary.

(3) From point *A*, the median line will run shoreward along the perpendicular bisector of line between salient shoreline points a and b to a point *B* which is equidistant from salient points *a* and *b* as well as from the next nearest salient point (*c*) on the coast of either state.

(4) From point *B*, the median line will run shoreward along the perpendicular bisector of a line between salient points c and b.

(5) This process is continued in this manner, with the median line running shoreward along the perpendicular bisectors of lines between any salient points, until the coastal termination of the common land boundary is encountered, with the last course being other than a perpendicular bisector.

A simpler description of the development of such a line is to simply run perpendicular lines from the midpoints of lines between selected salient points on the coastline. By definition, any points along such lines would be equidistant from the salient points at the ends of the midpoint lines. The intersection of perpendicular bisector lines from each pair of salient points would then be the inflection points for the equidistant line.

When coastlines with different ownership are opposite one another, and the distance between them less than twice the width of the territorial sea, the most equitable maritime boundary between them is also a median line. For that application, the median line would follow the perpendicular bisector of lines drawn between pairs of points on opposite side of the channel. Figure 6.2 in Chapter 6 (Boundaries for Littoral and Riparian Rights) illustrates such a line.

5.2 State Maritime Boundaries

Primarily based on the 1876 case of *Barney v. Keokuk* in which the U.S. Supreme Court had ruled that the public of each state held title to navigable inland nontidal waters, it was generally accepted that such title included the marginal sea bordering each states. Yet, in the 1930 with the discovery of oil off the California coast, disputes began regarding to the ownership of such waters. This led to a series of court cases which held that the federal government, rather than the states, owned the marginal sea. The resulting uproar led to the passage in 1953 of the Submerged Lands Act (*43 U.S.C., s1301-1 1970*) that conveyed submerged lands bordering the coastal states to the public of those states. That act defines the seaward limit of the public trust waters of each of the coastal states as follows:

> Section 4 – The seaward boundary of each original coastal state is hereby approved and confirmed as a line three geographical miles distant from its coastline or, in the case of the Great Lakes, to the international boundary. Any State admitted subsequent to the formation of the Union which has not already done so may extend its seaward boundaries to a line three geographical[2] miles distant from its coastline or to the International boundaries of the United States in the Great Lakes or any other body of water traversed by such boundaries. . . . Nothing in this section is to be construed as questioning or in any manner prejudicing the existence of any State's seaward boundary beyond three geographical miles if it was so provided by its constitution or laws prior to or at the time such State became a member of the Union, or if it has been heretofore approved by Congress.

---

2 A geographical mile is the length of one minute of arc on the equator or 6087.08 ft. A marine league is three geographical miles.

The question of whether any of the states were entitled, under the Act, to the exception for submerged lands greater than three geographical miles from the coastline was decided by the Supreme Court in 1960 (*U.S. v. Louisiana, Texas, Mississippi, Alabama, and Florida*). That decision held that Texas and Florida were entitled to submerged lands extending three leagues into the Gulf of Mexico due to the extent of their boundaries at the time of admission into the Union. The same decision held that Louisiana, Mississippi, and Alabama were entitled to a marginal sea of only three geographical miles.

As a result, the public trust lands of each of the coastal states are now considered to be the navigable interior waters of their state together with the portion of the territorial sea out to 3 mi from the mean lower low water line (even though the current policy of the federal government is to claim ownership out to 12 mi) except for the states of Florida and Texas with lands extending out to three leagues into the Gulf of Mexico.

The Submerged Lands Act provided a definition for a baseline from which to measure for the seaward boundaries of state water as follows:

The coastline for the purpose of measuring out to the seaward boundary is...... the line of ordinary low water along that portion of the coast which is in direct contact with the open sea and the line marking the seaward limit of inland water. (Section 2 (c), Submerged Lands Act)

The "line of ordinary low water" or baseline for measuring out to the boundary in the above definition is interpreted to be the mean lower low water line. The "seaward limit of inland waters" refers to any closing line drawn across bays, river mouths, or other inland waters. A more complete discussion of such closing lines may be found in Section 1.1.1 of this writing.

Since the 1960 Supreme Court ruling mentioned above, the court has continued to clarify some of the complex issues regarding state maritime boundaries. Such cases include *U.S. v. Florida* which defined where the Atlantic Ocean (with a 3-mi marginal sea) ended and the Gulf of Mexico (with its three-league marginal sea) began. In that case, the court found the Gulf of Mexico to be limited to areas north of latitude 24° 35′ north and west of longitude 83° west. As a result, the seaward boundary for Florida is three geographical miles from the coastline in the Atlantic and on the south side of the Florida Keys to a point at latitude 24° 35′ north. From that point, the boundary goes due west along the dividing between the Atlantic and the Gulf (latitude 24° 35′ north) to a point three marine leagues from the coastline. Then the boundary follows the Gulf coast at three leagues distance, northerly and westerly. The opinion established a separate zone, 3 mi wide to the south and three leagues wide to the north and west around the Dry Tortugas Keys (Figure 5.5).

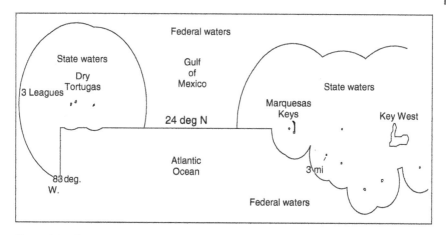

Figure 5.5 State/Federal Boundaries in the Florida Keys Area. *Source:* Cole (1997)/John Wiley & Sons.

Such cases also included *U.S. v. Louisiana* which, in 1985, determined the ownership of enclaves within Mississippi Sound which were located more than 3 mi from both the mainland shoreline as well as from the shoreline of the barrier islands bordering the sound. In that case, the court found that since Mississippi Sound was both a juridical bay as well as a historic bay,[3] the waters of Mississippi Sound were inland waters and entirely owned by the states. Prior to that decision, the question of whether such enclaves were state owned resulted in them being used as havens for near-shore locations for gambling operations prohibited under state law at that time.

References

Cole, G. (1997). *Water Boundaries*. New York: John Wiley and Sons.
Hodgson, R. and Alexander, L. (1972). *Towards an Objective Analysis of Special Circumstances, Occasional Paper 13*. Kingston, RI: Law of the Sea Institute.
Reagan, R. (1988). Territorial Sea of the United States of America, Proclamation 5928, Federal Register, 54 (WO5).
Shalowitz, A. (1962). *Shore and Sea Boundaries*. Washington, D.C.: U.S. Coast and Geodetic Survey.

---

3 See Section 5.1.1 regarding historic bays.

Case Law Cites

Barney v. Keokuk, 94 U.S. 324 1876.

U.S. v. California, 332 U.S. 19, 1947; 432 U.S. 40 1977.

U.S. v. Louisiana, 339 U.S. 699, 1950; 394 U.S. 1, 10 1969.

U.S. v. Louisiana, Texas, Mississippi, Alabama and Florida, 3 63 U.S. I. 212 1960.

U.S. v. Maine, 470 U.S. 515 1975.

6

Boundaries for Littorial and Riparian Rights in Public Trust Waters

6.1 Riparian and Littoral Rights

Riparian rights are rights of owners of upland bordering on water bodies related to the use of the water. The word, "riparian" rights (derived from the Latin "*ripa*," a river bank) actually applies only to lands bordering rivers and streams with "littoral" rights (derived from the Latin "*litus*," the seashore) applying to ocean front lands. Nevertheless, due to the general use of the word, "riparian" is used in this writing for all types of waters. These rights include a wide range of privileges associated with the adjacent water including the use of the water, the right to moor a boat, and often the right to wharf out into the water as well as the right to newly formed land caused by changes in the shoreline. All of these are provided that such use does not interfere substantially with the rights of others with such rights. The two issues regarding these rights to be addressed in this writing include the division of adjacent waters for the exercise of such riparian rights as the mooring of boats, the construction of docks into the water, and access to navigable waters; and the division of newly formed land caused by changes in the shoreline.

Riparian rights also apply to ownership or use of submerged land lying beneath non-sovereign waters. As a result, the same procedures for division of such privately owned submerged lands as those for determining the limits for ownership as for riparian rights over adjacent waters. See the following section for discussion of those procedures.

6.2 Riparian Rights in Adjacent Waters

Riparian rights over adjacent waters include the right of access to and reasonable use of the water provided that it does not interfere substantially with the rights of others with such rights. These rights include the access to navigable waters, the

Sea Levels and Coastal Boundaries, First Edition. George M. Cole.
© 2024 John Wiley & Sons, Inc. Published 2024 by John Wiley & Sons, Inc.

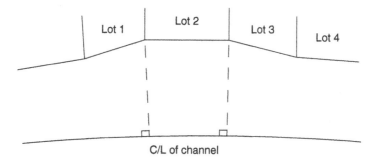

Figure 6.1 Division of Area of Riparian Rights in Rivers.

right to moor a boat, and often the right to wharf out into the water. As previously noted, the procedures described for division of submerged lands for riparian rights in adjacent waters also apply to the division of submerged lands beneath non-sovereign waters which are owned by the adjacent upland owners.

Determining the division line between the riparian rights areas for adjoining land owners can be challenging since acceptable approaches vary considerably with the configuration of the shoreline and the upland property boundaries. This variability has been noted judicially with statements such as "No geometric theorem can be formulated to govern all cases" (*Hayes v. Bowman*). While that statement is certainly true, there are some general rules of procedure.

There are three basic guidelines when dealing with the area within which such rights apply. (1) The length of the affected tract's water frontage, not its total acreage or upland lot lines, generally controls the apportionment of riparian rights. (2) Points of departure for extending dividing lines into the water should be at the intersection of the upland lateral boundary and the actual shoreline as opposed to lines shown on subdivision plats or the meander line of government surveys. (3) The proper procedure for this does not involve extension of the upland lateral boundary without change of direction.

The method for determining the proper direction for projection of division lines varies with the nature of the water body. In rivers and streams which are significantly narrow to determine a *thread,* or center of stream, the prevailing procedure is to project the line in a direction perpendicular to the thread of the stream[1] (Figure 6.1).

Where the geometric center of the stream is considerably removed from the *thalweg,* or deepest part (channel) of the stream, there may be justification for

---

1 Typical case law suggesting this approach includes *Knight v. Wilder, Clark v. Campau, and Wood v.* Appal.

using the thalweg as the center. Regarding this distinction, one source (Bade 1940) defines the thread as follows:

> The term 'thread of the stream' means the geographic center of the stream at ordinary or medium stage of the water, disregarding slight and exceptional irregularities in the banks. It is fixed without regard to the main channel of the stream. . . . . . If the stream is made a boundary in a private conveyance, then the thread of the stream will be the stream boundary.

Regarding the center of the stream as being the "thalweg" or deepest part of the channel, *Stubblefield v. Osborn* addresses that issue as follows:

> Upon principle, therefore, it would appear that the thread of a nonnavigable river is the line of water at its lowest stage. The thread or center of a channel, as the term is above employed, must be the line which would give to the landowners on either side access to the water, whatever its stage might be and particularly at its lowest stage.

Ideally, when the thread is being used, a median line should be constructed to define it using field surveyed or scaled coordinates for salient points along the two sides of the shoreline. Although somewhat complex in construction, a median line (Bureau of Land Management 2009) provides a more precise and defendable baseline than as opposed to an approximate line.

To create a median line, a series of salient shoreline points should be selected at roughly equal distances along both sides of the channel. The median line passes through mid-points of the lines between each set of opposing points and is perpendicular to each such line. Inflection points on the median line are at the intersections of segments of the line passing through the mid-points (Figure 6.2). Once coordinates for the salient points are determined, coordinates for inflection points on the median line may be easily determined using a CADD program. Alternately, the coordinates may be calculated using coordinate geometry.

With oceans, large bays, or wide rivers, the thread is typically not readily discernible. In such situations, case law typically suggests using lines run perpendicular to various other baselines such as the shoreline if it is relatively straight, the bulkhead or pier line, or the channel line.[2]

---

2 Typical cases suggesting this approach are as follows: FL (In direction of channel) *Hayes v. Bowman*; LA (Perpendicular to shoreline) *Municipality No. 2 v. Municipality No. 1*; RI (perpendicular to harbor line) *Manchester v. Point St. Iron Works*.

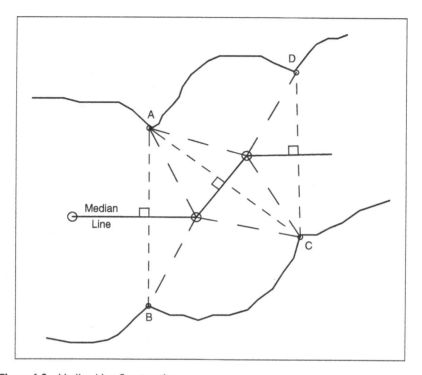

Figure 6.2 Median Line Construction.

Where the shoreline in question is curving or irregular such as in coves, it has been generally held that the proper procedure is to give each riparian tract a proportionate share based on measurement of the shoreline in question as well as measurement of some outer line such as a channel line or harbor line[3] (Figure 6.3).

For lakes that have a round shape, the general rule holds that a point at the geographical center of the lake be established by partition lines running from that point to the ends of the upland lateral boundaries (Markusen v. Mortensen). For lakes with significant variation in depth, a center point in the deepest part of the lake should be used to provide equal access to deeper waters (Figure 6.4).

For linear or irregular shaped lakes, a centerline running the length of the lake should be constructed and the same principles should be used as for streams. At the end points of the center line, converging lines similar to those used for a round

3 Cases illustrating this approach are as follows (Foster 1959): MA (*Deerfield v. Arms*); MD (*Baltimore v. Baltimore & P.S.B. Co.*); VA (*Groner v. Foster; Waverly Waterfront & Improvement Co. v. White; Cordovana v. Vipond; Rice v. Standard Products Co.*)

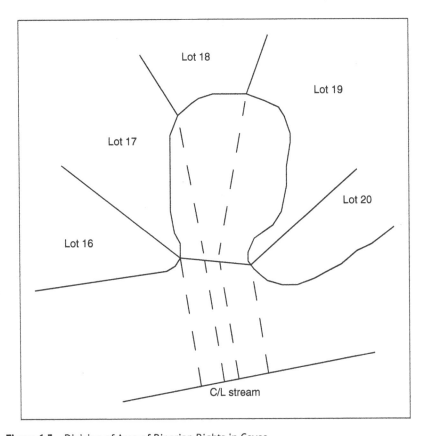

Figure 6.3 Division of Area of Riparian Rights in Coves.

lake should be used (Figure 6.5). This approach was stated clearly in case law as follows (Hardin *v.* Jordan):

> Where a lake is very long in comparison with its width, the method applied in rivers and streams would probably be the most suitable for adjusting riparian rights in the lake bottom along its sides and the use of converging lines would only be required at its two ends.

Although the method for determining limits of the riparian rights area associated with an upland tract varies significantly with the situation, the objective is to apportion the water area over which riparian rights are available in proportion to the relative length of each riparian tract. As a result, the length of a riparian parcel's shoreland, not its total area or the direction of the upland lateral lot lines, generally controls the apportionment of riparian rights.

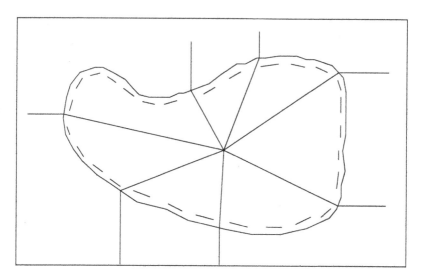

Figure 6.4 Division of Area of Riparian Rights in Round Lakes (Pie Slice Method).

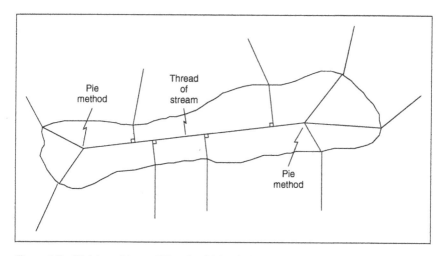

Figure 6.5 Division of Area of Riparian Rights in Long Lake.

From the above, it may be seen that there are at least four basic cases – narrow streams, wide water bodies, irregular shorelines, and lakes – with each case having its own rules for the division of riparian rights. With narrow rivers, the general rule is to use a division line perpendicular to the thread of the stream. For wider water bodies, the general rule is to use a line run perpendicular to the shoreline or to some other baseline such as a bulkhead or channel line. For lakes, the general rule is to use lines drawn to a common center point or center line. For curving or irregular shorelines, proportional division of a closing line based on length of the shoreline is the general rule.

For more complex situations requiring the division of riparian rights in adjacent waters, a proportionality test may be used to evaluate the fairness of a proposed apportionment. That technique compares the ratio of the areas allotted to two adjacent riparian lots to the ratios of the lengths of their respective shorelines. Such an evaluation is based on methods used to evaluate proposed division of continental shelf areas for international boundaries. If the ratios are not similar, the proposed division may be inequitable.

Approaches other than those described above are necessary for certain applications such as the *colonial method* used for division of tidal flats in some of the New England states (Figure 6.6). That method uses proportionate measurement between the low and high water marks. Instructions for the use of that method as prescribed in the case of *Emerson v. Taylor* follow:

> Draw a base line from the two corners of each lot, where they strike the shore, and from those two corners, extend parallel lines to low-water mark, at right angles with the base line. If the line of the shore be straight there will be no interference in running the parallel lines. If the flats lie in a cove, of a regular or irregular curvature, there will be an interference in running such lines, and the loss occasioned by it must be equally borne or gain enjoyed equally by the contiguous owners. . .

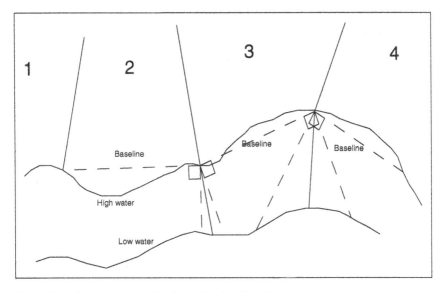

Figure 6.6 Colonial Method for Proportioning Tidal Flats.

Referring to Section 3.1.2 of this writing, some of the New England states have adopted mean low water as their coastal boundary. As a result, the colonial method is for determining ownership rights over tidal flats as opposed to usage rights as with other riparian or littoral rights. Nevertheless, it is included in this section since that can certainly be considered a littoral right.

6.3   Riparian Rights to Newly Formed Land

Coastal land is typically in a constant state of flux. In some areas, new upland is being formed by material deposited along the shoreline (*accretion*) or by decline in water level (*reliction*). In other areas, upland is being lost due to *erosion* or by rises in water level.

The general rule is that the upland owner gains title to new upland created by reliction and accretion and loses title to lands submerged by rises in water level or lost by erosion. Yet, when such changes are not gradual or imperceptible, or where such changes are artificially induced, the general rule may not apply. For sudden changes, such as those occurring during storms, the process is termed *avulsion*, and it has been generally held that title does not change. Likewise, it has been held that shoreline changes resulting from man-made actions, such those associated with dredging or groins, do not result in change of ownership if the upland owner or a predecessor in title caused the changes. This is the case in the State of Florida where prevailing judicial opinions have held that artificial accretion caused by the upland owner remains the property of the sovereign (*McDowell v. Trustees of the Internal Improvement Trust Fund of State of Florida*) while artificial accretion caused by third parties accrues to the upland owner (*Board of Trustees v. Madeira Beach Nominee Inc.; Board of Trustees v. Sand Key Associates Ltd*).

When shoreline changes do result in a title change, the division line across the new lands between the adjacent riparian owners may have to be determined. Generally, similar procedures to those for riparian rights in adjacent waters are used for that purpose.

The general rule regarding ownership of accreted lands also applies to islands that emerge in water bodies. It is generally held that islands formed in a water body belong to the owner of the bed of water body. If an island forms in state sovereign waters, it is the property of the public of that state. If the riparian proprietors own the bed of the waters, an emergent island belongs to the proprietor on whose side of the stream it formed. If an island is formed in the middle of a stream owned by the riparian proprietors, the island is divided by the line along the thread of the stream. In public land states, islands that were existent at the time of the original public land survey but were overlooked in the survey are considered to be *omitted lands* that remain in the ownership of the federal government until validly conveyed by patent.

6.4 Dealing with Avulsion or Altered Shorelines

As mentioned in the previous section, when shoreline changes are due to avulsion or actions of the littoral owner, it has generally been held that title does not change. Examples of such situations include sudden storm damage, artificial, accretion caused by actions of the littoral owner, dredging of man-made channels, use of artificial fill to create upland or enhanced docking facilities, municipal beach renourishment projects, creation of man-made harbors, and ditching for mosquito control.

When determining water boundaries in such cases, it may be necessary to locate the last natural position of the shoreline. The best evidence of the pre-alteration shoreline typically is a survey plat or map which specifically delineates the desired water boundary. When such evidence is not available, other sources must be considered, such as the following.

T-Sheets: For major coastal water bodies, some of the better sources of historic shoreline locations are shoreline topographic maps produced by the U.S. Coast Survey, presently a component of the National Ocean and Atmospheric Administration (NOAA). These maps, commonly called "T sheets," are the original field survey manuscripts prepared for use in compiling nautical charts. Such maps have significant detail, generally specifically mapped the mean high water line, and are georeferenced (Figure 6.7). The first T-sheet was produced in 1834. Most of the coastline for the conterminous United States was mapped between that time and the late 1800s. Until the late 1930s, these maps were created by plane table surveys and have subsequently been created by photogrammetry using aerial photographs. No field notes were produced in the plane table surveys since angles were measured graphically and distances determined optically by stadia. As a result, all work is shown on the manuscripts themselves.

On such maps, the high water line is the most prominent feature since it represents the line between land and water on the subsequently produced nautical charts. The earliest specific instructions regarding the nature of the shoreline to be located were developed in 1898. Those instructions called for the location of the *"average high-water line."* The intent was apparently always that of the mean high water line. These maps also depict the mean lower low water line and various other features, both natural and man-made, observed by the survey team. Exceptions to the depicted shoreline on T sheets representing the mean high

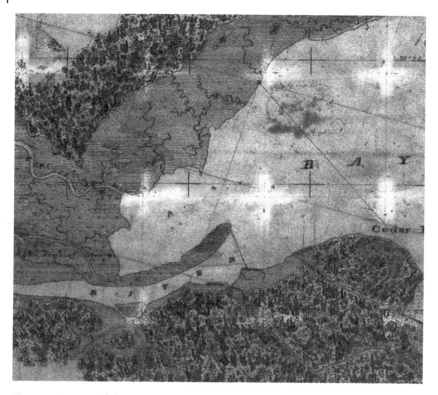

Figure 6.7 Typical "T" Sheet.

water line occur in areas of tidal marshes. In such areas, the seaward edge of the marsh was mapped as opposed to the mean high water line (Shalowitz 1962; Swanson 1982) since that line was the one observable to mariners. Care must be taken in interpreting such maps since the same symbol was used to represent the outer edge of marsh line as well as the mean high water line on most early surveys (Shalowitz 1962). On more recent shoreline manuscripts, the outer edge of marsh is usually labeled as the "apparent shoreline" and was indicated with a fine line as opposed to the standard weight line used to indicate the mean high water line. When marshes or grassy flats which were generally flooded at high water were encountered, these were typically indicated on the map with a marsh symbol without a bordering line. In addition, it is noted that the inner or upland edge of marsh was shown by a line on many of the earlier surveys, although this practice was discontinued on later surveys. Such a feature was not intended to represent the mean high water line and should not be interpreted as such.

An advantage of using such maps is that the data contained on them are georeferenced. As a result, that data can be compared with contemporary surveys.

Nevertheless, since there have been several changes in the reference spheroid and in the national horizontal datum over the years, the relationship between the system with which the map was compiled and the current system, the relationship between the two systems used must be known. Generally, hand corrections were made to early T sheets to allow for changes in geographic systems. This was usually done by adding corrective grid ticks adjacent to the original ticks (Figure 6.8).

Other Coast Survey Products: In addition to T-sheets, the U.S. Coast Survey and its successor agencies have produced a number of other products which may be of use in determining historic shoreline locations. These include hydrographic survey sheets and descriptive reports. Hydrographic survey sheets are field manuscripts containing plots of the various hydrographic soundings and were produced for nautical chart production. Shorelines depicted on hydrographic sheets were usually traced from the accompanying T-sheets. Descriptive reports were prepared by the officers directing the field work for the purpose of aiding the cartographer in

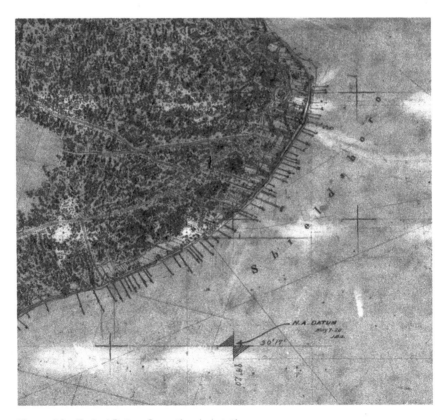

Figure 6.8 Typical Datum Correction Annotation.

preparing the final nautical chart. Such reports contain information helpful in interpreting and evaluating the accuracy of the topographic and hydrographic surveys. Information on the sources of the tidal data used in the project is often included.

Public Land Surveys: The Bureau of Land Management of the U. S. Department of the Interior and its predecessor agency, the General Land Office (GLO), has had the responsibility for surveying the public lands of the United States. Because a large percentage of the land within the country passed through federal ownership, those surveys represent the earliest systematic surveys in many areas of the country. Such surveys were for the purpose of subdividing federal territory into 6-mi square townships and then into 1-mi square sections for subsequent conveyance to private ownership. With such surveys, work products are field notes and plats prepared from the notes.

When navigable water bodies were encountered, the public land surveyors generally ran meander lines, or surveys of the margin, of the water body. The purpose of such meander lines was to segregate the submerged lands which were reserved as highways of commerce from land that was to be conveyed to private ownership. Meander lines were not intended to be precise mean high water line or ordinary high water line surveys. The meander points themselves may have been correctly located on the high water line, but they were infrequently located. Meander lines were intended to provide only a general representation of the shoreline. Further, the early public land surveys were not georeferenced and generally were not as specific in identifying the high water line. Yet, with an understanding of their limitations, these surveys may be a useful source of historic shoreline location.

To aid in the interpretation of these surveys and weighing their significance, it is helpful to examine the instructions issued by the GLO to the deputy surveyors performing such surveys. The earliest of such instructions, issued by Surveyor General Tiffin in 1815, merely states that *"for meandering rivers you will take the bearings according to the true meridian of the river and note the distance on any course when the river intersects the sectional lines. . . . . ."* Subsequent instructions were generally more specific. As a result, when using these surveys, it is prudent to also review the instructions under which they were prepared.

Aerial Photography: For periods after the 1930s, repetitive aerial photography may be available to assist in determining pre-alteration shorelines. For making precise measurements from aerial photography, various steps are necessary. These include the interpretation of the photography to identify the water boundary and the rectification of the photo imagery to a current horizontal datum. Sources of systematic aerial photography in the federal government include the National Ocean Service and its predecessor agencies, the U.S. Geological Survey, the Soil

Conservation Service of the U.S. Department of Agriculture, and the National Aeronautics and Space Administration. In addition, aerial photography may often be obtained from various state agencies such as transportation and revenue departments and private photogrammetric firms.

Other Data Sources: In addition to these specific sources, historic maps are often available from various other sources. These include the National Archives, the U.S. Corps of Engineers, the U.S. Geological Survey, various state land offices, county court records, private surveying firms, title companies, and local historic societies. Depending on the site, other evidence may be available for such determinations. This includes maps from various sources, testimony of eye-witnesses, soil analysis, and vegetation analysis.

References

Bade, E. (1940). Titles, points, and lines in lakes and streams. *Minnesota Law Review* 24: 305.

Bureau of Land Management (2009). *Manual of Instruction for the Survey of the Public Lands of the United States*. Washington, D.C.: U.S. Government Printing Office.

Foster, C. (1959). *Annotation on Apportionment and Division of Area of River as Between Riparian Tracts Fronting on Same Bank in Absence of Agreement or Specification*, 65 American Law Reports 2nd 143.

Shalowitz, A. (1962). Shore and Sea Boundaries, Pub. No. 10-1, U.S. Coast and Geodetic Survey, Washington, D.C.

Swanson, L. (1982). Personal communication.

Case Law Cites

Baltimore v. Baltimore & P.S.B. Co.

Board of Trustees v. Madeira Beach Nominee Inc., Fla., 272 So. 2d. 209 (1973).

Board of Trustees v. Sand Key Associates Ltd., 512 So. 2d. 209 (1973).

Cordovana v. Vipond, Va, 94 SE 2d. 295 (1956).

Deerfield v. Arms, 28 Am Dec 27. Mass (1835).

Emerson v. Taylor, 9 Me. 3 (1832).

Groner v. Foster, 27 SE 493, Va (1897).

Hardin v. Jordan, 140 U.S. 371 (1891).

Hayes v. Bowman, Fla., 91 So. 2d. 795 (1957).

Manchester v. Point St. Iron Works.,13 R.I. 355 (1881).

Markusen v. Mortensen, Minn., 116 NW 1021 (1908).

McDowell v. Trustees of the Internal Improvement Trust Fund, 90 So. 2d. 715 (1956).

Municipality No. 2 v. Municipality No. 1, 17 LA 573 (1841).

Rice v. Standard Products Co., 99 SE 2d. 529, Va (1957).

Stubblefield v. Osborn, Neb.,31 NW 2d. 547, 1948.

Waverly Waterfront & Improvement Co. v. White, 33 SE 534, Va (1899).

7

Sea Levels and Coastal Boundaries Yesterday–Today–Tomorrow

7.1 Sea Levels and Coastal Boundaries – Yesterday

Based on this writing, it is hoped that readers have taken away an understanding that sea levels have been in a state of flux throughout the history of the Earth. Further, if the past is indeed prologue, that status will continue. Developments in geological science have gradually provided an understanding of historic sea level patterns associated with past global glaciation cycles as discussed in Chapter 1. As a result, it is now recognized that, for at least the last half-million years or so, the Earth has gone through repeated glacial/interglacial cycles with periods generally ranging from approximately 50 to 100 thousand years. The last glacial period is believed to have begun about 130,000 years ago and ended about 20,000 years ago. (Interestingly, the last glacial period may have permitted migration of Asians to North America via a land bridge across the Bering Strait (Emery and Aubrey 1991).)

At the end of the last glacial period, levels in the sea began rising at rapid rates due to the melting of glaciers. Those rates gradually stabilized and for the last six thousand years or so, sea levels worldwide have generally risen more slowly. Although humans evolved well before the current interglacial period, many of the rudiments of current human civilization did not appear until sea levels more or less stabilized about 6000 years ago (Day et al. 2023). As a result, current human society is primarily a product of the portion of the interglacial period with stable or slightly rising sea levels. Once sea level did stabilize, many early humans were attracted to the seas and other great waters of the Earth (Cole et al. 2017). That attraction was due to a number of factors including the role of those waters as a source of abundant food, moderate climate, and as an avenue of transportation. In most coastal areas, the adjacent alluvial plains provided flat and fertile land conducive to agriculture, and the mild climate found near the coast made life is more comfortable. In addition, such areas provided access to the sea for trade and

Sea Levels and Coastal Boundaries, First Edition. George M. Cole.
© 2024 John Wiley & Sons, Inc. Published 2024 by John Wiley & Sons, Inc.

commerce. Because of such factors, coastal areas were the location of some of the earliest human settlements.

As improvements in ships, navigation, and technology developed, humans began to increasingly use the waters of the Earth. That resulted in many of the world's largest cities being located in coastal areas. Coastal communities became important centers of worldwide exploration, trade, and the fishing industry. The intense coastal development was true not only along the seashore, but also along the coasts of major bays, rivers and lakes. As more affluent societies developed, additional people have been drawn to such coastal areas due to the mild climate, scenic views and recreation opportunities. A large percentage of the human population now lives in coastal zones. As an illustration of this, recent United States census data indicates that at least fifty percent of the nation's population reside in counties bordering the coasts or major arms of the sea. As a result, a great deal of coastal land has become intensely developed and is now among the most valuable lands in many areas.

Due to the attraction for humans to coastal areas, policies designating the great waters of the Earth as public commons developed early in current society. Along with that designation and the popularity of coastal areas, definitions of the boundaries between the waters and the land became standardized. In addition, definitions were developed by the Law of the Sea Treaty of 1982 for standard widths of national territorial seas. Both the standardization of coastal boundaries and national claims to territorial seas have allowed more stable use of coastal lands and the bordering waters in developed society.

Since most coastal developments are the products of that portion of the current interglacial period with stable or moderate rates of sea level change, many areas have been developed below levels reached in earlier interglacial periods. Such areas may be threatened with inundation if sea levels rise much above current levels. Further, some coastal areas, especially those along the Atlantic and Gulf coasts of the United States, have recently experienced apparently higher rates of sea level rise. This has resulted in considerable public concern. Many researchers have attributed such higher rates to human-caused global warming. Nevertheless, very recent studies have suggested such increases to be, at least partially, associated with sinking of the land related to natural tectonic forces that cause vertical land motion or human activities such as excessive pumping of groundwater, extraction of underground minerals, and extensive development (Seo et al. 2023; Ohenhen et al. 2023). Interestingly, that is not a new conclusion. That same view was expressed very clearly with the following statement from two well-qualified coastal geologists three decades ago as follows:

> "Our analysis shows that the signal of a possible eustatic rise of sea level is obscured by "noise" caused mainly by movement of the land beneath tide gauges; thus, study of the "noise" is a potential source of information about

modern movements of the Earth's crust – especially of plate tectonics. These data support concepts from geological and geophysical observations of sinking or rising land levels related to crustal movements associated with glaciation, subduction, rifting, sediment loading, volcanism, and general faulting and folding associated with the properties and history of coastal plates and especially their movements during sea-floor spreading. This conclusion should be no surprise to geologists, but it may be unexpected by those climatologists and laymen who have been biased too strongly by the public's perception of the greenhouse effect on the environment. At present, we cannot discover a statistically reliable rate for eustatic rise of sea level alone, but it may not matter to the seafront property owner whether his house becomes flooded because of a rising sea or a sinking coast.

The present rate of relative sinking of land levels along coasts is less than a tenth of that during the time of fastest melt and retreat of late Pleistocene ice sheets, and more than 10 times the average for the past 50 m.y., during most of which the effects of glaciation were absent and geological agents along were effective. In addition, many present coastal changes (erosion of beaches, collapse of sea cliffs, disappearance of salt marshes, sinking of some coastal cities, siltation of harbors, and others) are caused more by indirect and unexpected results of human activities than to changes of sea level or land level, Most coastal instability can be attributed to tectonism and documented human activities without invoking the spectre of greenhouse warming climate or collapse of continental ice sheets." (Emery and Aubrey 1991)

7.2 Sea Levels and Coastal Boundaries – Today

7.2.1 Long-Term Trends

Although at much slower rates than immediately after the last glacial period, sea levels have been steadily rising since that time. Due to intense development in coastal zones, there is currently considerable public interest in sea level trends. To address that interest, networks of water level (tide) gauges have been established along the coastlines of the United States as well as in developed coastal countries worldwide. Those networks assist in monitoring sea level change. Yet, it is now recognized that changes in ground-based water level measurements, such as those produced by water level gauges, do not necessarily reflect actual sea level rise. As mentioned previously, vertical land motion due to various natural, as well as human-caused, processes can be significant, and sometimes dominant, factors

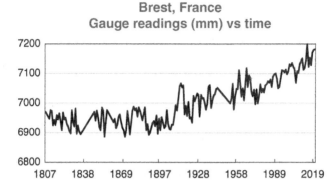

Figure 7.1 Water Level Gauge Readings for Brest, France.

in such measurements. Hopefully, future sea level measurements will be accompanied by satellite observations, such as those available with global positioning systems (GPS) to allow a true picture of what is happening along coastlines today.[1] Such satellite observations may allow detection of changes in ground level that affect the gauge readings. Thus, this will allow correction of water level gauge readings to allow determination of actual sea level change rates. This may also allow identification of changes in land use practices needed to minimize such vertical land motion if they are human caused. Such satellite observations need not be continuous. Rather, annual satellite observations at permanent water level monitoring stations would serve to determine nearby trends in vertical land motion. Therefore, such additions to sea level monitoring networks need not be a costly improvement and are strongly encouraged.

Despite the limitations of ground-based sea level measurements, such as those provided by water level monitoring gauges, they provide the only systemically collected data for monitoring sea level change. Therefore, as part of the research conducted for this writing, an informal analysis was conducted of measurements from several key water level gauges. For these data, annual sea level values from the Permanent Service for Mean Sea Level were used although the original source of the U.S. data was the National Oceanic and Atmospheric Administration (NOAA). As a first step in that research, average annual measurements from the longest continually operating gauge in the world, Brest, France, which began observations in 1807, were examined (Figure 7.1). As may be seen from that figure, the slope of the measurements from that station is complex and best fits a high-order polynomial curve. Beginning with a downward trend, which was

---

1 Currently the NOAA National Geodetic Survey does maintain a series of continuous operating GPS stations. Yet, with few exceptions, they are not co-located with NOAA tide stations.

ongoing when observations began in 1807, sea level as measured by that gauge began a long linear rise in the mid-1800s. It then began to slope slightly downward for a brief period in the mid-twentieth century before beginning to rise again in the later portion of that century.

To examine the sea level trends along the United States coastlines, annual average sea level measurements from long-operating water level stations in the United States were examined for a 120-year period (1900–2020) along with data from the Brest station as a baseline. As representative of the U.S. Pacific coast, data from the longest operating tide station in the nation, San Francisco, California (installed in 1850), was used. As representative of the U.S. Atlantic coast, data from the second longest operating water level station in the United States, the Battery in New York City (installed in 1851) was used. The average linear trends for each year over the entire period for all three stations were determined, and the resulting measurements are plotted to provide a visual look at the trends (Figure 7.2).

For the data in Figure 7.2, the linear least-squares trend rate for each of the stations for the entire 120-year period was also calculated. That exercise resulted in trends as follows:

Brest, France	1.5 mm/yr
San Francisco, California	1.9 mm/yr
New York Battery, NY	3.1 mm/yr

7.2.2 Recent Trends in Sea Level

In addition to examining long-term trends, an analysis of the trends of sea level change for a recent 19-year tidal epoch (2002–2020) for an extended selection of

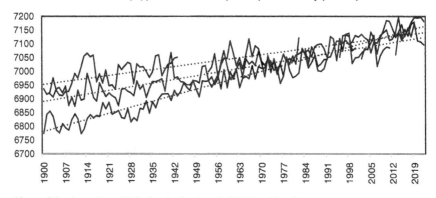

Figure 7.2 Long-Term Variation in Sea Level, 120-Year Trends.

water level stations for the Atlantic, Gulf, and Pacific coastlines of the United States was made with results provided in Table 7.1. For comparison purposes, the trend for the Brest, France station is included. Although there are a limited number of non-US water level stations with continuous observations without

Table 7.1 Rate of Sea Level Change (mm/yr), for 2002–2020 Tidal Epoch.

Station	Trend (mm/yr)
Europe	
Brest, France	3.5
Atlantic Coast	
Boston, MA	3.6
Battery, NY	5.3
Atlantic City, NJ	5.6
Lewes, DE	6.9
Sewells Pt., VA	7.7
Beaufort, NC	8.7
Myrtle Beach, SC	8.6
Charleston, SC	10.1
Savannah, GA	10.2
Port Canaveral, FL	8.6
Virginia Key, FL	8.2
Gulf Coast	
Key West, FL	7.9
Clearwater, FL	9.2
Cedar Key, FL	9.9
Apalachicola, FL	8.8
Dauphine Is., AL	11.9
Grand Is., LA	11.8
Galveston, TX	13.8
Pacific Coast	
Los Angeles, CA	4.4
San Francisco, CA	4.6
South Beach, OR	1.6
Neah Bay, WA	−1.0
Juneau, AK	−15.5

significant data gaps, the results for that station appear to be representative of typical station in the temperate zone of the northern hemisphere. As examples, the following rates for the same epoch were found in other areas: Portsmouth, UK: 3.2 mm/yr, Sines, Portugal: 3.9 mm/yr, Le Couquet, France: 3.4 mm/yr, and Kahului, Maui, Hawaii: 3.8 mm/yr.

As may be seen from Table 7.1 results, interesting geographic trends were observed along all three of the U.S. coastlines. The wide disparity of those data was astonishing high. The Gulf Coast data clearly supports the long-recognized theory that portions of that coast were sinking due to extraction of minerals or groundwater. The trends along the U.S. Atlantic for the 2000–2022 period appear to support results from recent studies (Ohenhen et al. 2023) that found portions of the U.S. Atlantic coast to also be "sinking". When recent rates of sea level change for the Atlantic Coast from Table 7.1 are plotted versus approximate distance along the coastline (Figure 7.3), there appears to be a definite geographic trend with a highpoint in the vicinity of Savanah, Georgia and Charleston, South Carolina. For most open coastlines, the rate of change would be expected to be more similar for actual sea level changes. Therefore, this trend suggests that the cause of the varying rates is most probably vertical land motion rather than changes in the rate of actual sea level rise.

Since Charleston, South Carolina appears to have one of the higher sea level increase rates among the stations examined along the Atlantic coast, the trends for that station were examined in greater detail. First, the trend for that station over a recent 19-year tidal epoch was plotted (Figure 7.4). As may be seen from that figure, the trend fits a third-order polynomial curve with a 0.85 coefficient of

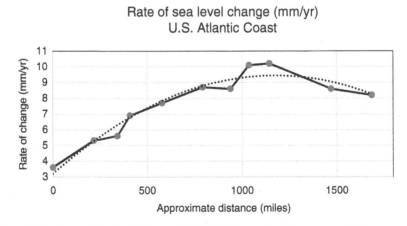

Figure 7.3 Rates of Sea Level Change (2002–2020 epoch) v. Approximate Distance, along U.S. Atlantic Coastline.

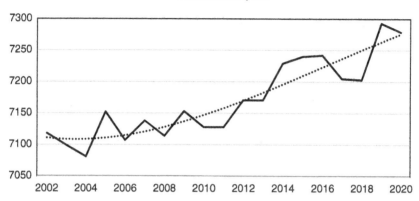

Figure 7.4 Variation in Sea Level (mm) for Recent Tidal Epoch, Charleston, SC.

determination (R^2). The trend line for those data clearly reflects a definite rising trend with a linear rate of rise for that period of 10.0 mm/yr.

The trend for that station over a longer period of 100 years was also examined (Figure 7.5). As with the shorter period trend, the trend curve for that figure clearly reflects an upward trend in sea level for a decade or two ongoing when observations began in 1922 and then another beginning about in the 1980s. An analysis indicates an average rising trend of 3.5 mm/yr over the 100-year period.

In addition, the long-term trend for that station was compared with that of relatively nearby Fernandina, Florida (Figure 7.6). As may be seen, the two stations demonstrated very similar patterns and trends. The Fernandina station, with a

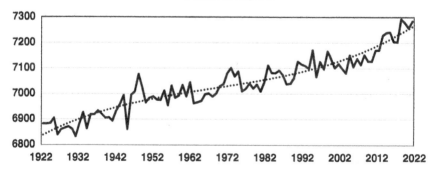

Figure 7.5 Variation in Sea Level (mm) for 100 years, Charleston SC.

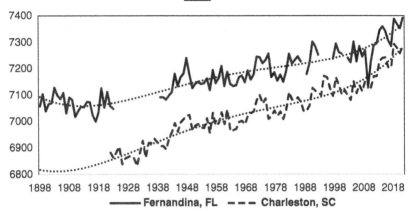

Figure 7.6 Long-Term Variation in Sea Level (mm), Charleston SC and Fernandina, FL.

longer period of operation, had a downward trend in the early 1900s, then both stations showing upward trends starting roughly in the 1920s and then again in the 1980s. For the entire 1940–2022 period, Charleston had an average rise rate of 3.4 mm/yr, while Fernandina had an average rate of 2.5 mm/yr.

In addition, the **rate of change** for sea level was examined for Charleston (Figure 7.7). That involved calculating the differences between the 19-year

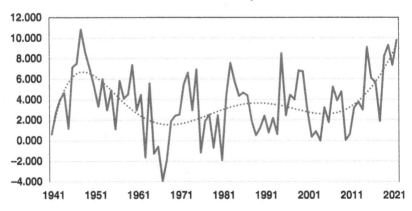

Figure 7.7 Sea Level Change Rate, Charleston SC.

average for each year of record and that of the previous year for the last 80 years. Interestingly, that analysis suggested that change occurred in surges or waves. No explanation has been found for this phenomenon although the same pattern appeared to also exist in other tide stations along the southern Atlantic coast of the United States. It may be that the pattern is related to meteorological trends, but further research is indicated.

7.2.3 Variations in Rate of Change

In addition to examining the long-term trends in sea level and the analysis of the Charleston tide station record, a study was made of trends indicated by available records for other key water level stations along the Atlantic, Pacific, Gulf, and South-East Alaskan coastlines of the United States for a series of tidal epochs over the twentieth and current century. The results of that analysis are provided in Table 7.2 and graphically in Figure 7.8. As may be seen, the rates of apparent sea level change for the U.S. coastline were significantly different than that of the Brest station which is on the European Atlantic coast.

As may be seen from the above results, sea level trends at the Brest, France station as well as those for San Francisco were within expected limits based on recent studies. Yet, trends for Charleston and Grand Island were considerably higher than earlier analyses. **The variation in trends for the U.S. Atlantic Coast suggests that the increase is probably not caused by actual sea level change but vertical land motion.** The wide disparity in those data is astonishingly high and appears to support results from a recent study (Ohenhen et al. 2023) that found portions of the East Coast of the U.S. to be "sinking". While that study suggested that such sinking may possibly be human caused, the relatively abrupt changes in trends indicated by the above data, especially those provided in

Table 7.2 Sea Level Rates of Change (mm/yr) (Bold Text Indicates Higher Rates of Change).

	Brest, France	San Francisco	Charleston, SC	Grand Is, LA	Yakutat, AK
1900–1919	—	1.0	—	—	—
1920–1939	2.4	2.9	—	—	—
1940–1959	—	1.3	3.3	—	—
1960–1979	—	1.2	3.2	—	**−16.8**
1980–1999	2.3	2.8	**4.8**	**8.0**	**−18.4**
2000–2019	2.7	3.0[a]	**8.2**	**11.8**	**−22.1**

[a] Period of 2000–2022 used.

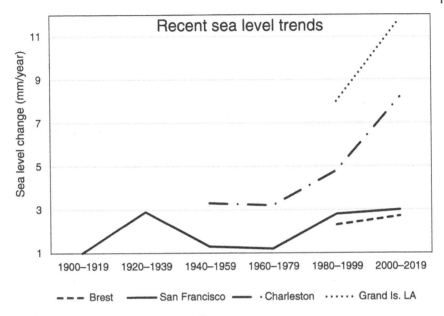

Figure 7.8 Recent Sea Level Rates of Change.

Table 7.1, suggest that tectonic motion is more likely the cause. **Although additional studies are warranted, the differences between trends along the U.S. coasts and that for Brest suggest that vertical land movement may be a larger factor than actual changes in sea level in concerns about our shorelines.** While there has long been a consensus that the wild variation in apparent sea level trends for the Western Gulf of Mexico and Southeast Alaska are related to vertical land motion, this also seems to be the case for the U.S. Atlantic coast. As suggested by the quote from Emery and Aubrey which was provided in Section 1 of this chapter, it may not matter to the seafront property owner whose house is being flooded. Indeed, the results are the same. Nevertheless, it is important to know the cause. If the causes of the vertical land motion are determined to be related to human activities, remedial action may be possible. As a result, routine satellite observations at key water level monitoring stations are important to assist in determining the cause of the change.

   Based on the analysis of water level data reflected in this section, it is concluded that worldwide sea levels appear to currently be rising at the rate of 3–4 mm per year. Nevertheless, the Atlantic, Gulf, and Northern Pacific coastlines of the United States are experiencing considerably different rates due to vertical land motion. A similar conclusion may be obtained from graphics available on the website of the NOAA Laboratory for Satellite Altimetry.

7.2.4 Impact on Coastal Boundaries

The current apparently rapidly changing sea levels along the coastlines of the United States have significant impacts on coastal boundary determination. The most important impact is that any determination of a coastal boundary should be associated with a specific date of survey and identification of the epoch for the datum.

Another important associated issue for coastal boundary surveys considering the apparently more rapidly changing sea level is the epoch for the data used in the survey. Currently, NOAA has a policy of consideration of changing the National Tidal Datum Epoch every 20–25 years. As mentioned in Chapter 1, due to the more rapidly changing apparent rate of sea levels in areas such as Southeast Alaska and the northwestern Gulf of Mexico, a five-year computational period has been adopted in lieu of the 19-year period for such areas. With the current apparently rapid change in sea level along the Atlantic coast, it appears as if a similar policy should be considered for all of the national coastlines. If that is too much of a burden for the agency, having the basic data and recommendations for the approach readily available for private surveyors to use might be a more economical approach to this problem. For either case, it is important that the epoch or computational period for the data used be identified on coastal boundary surveys.

7.3 Sea Levels and Coastal Boundaries – Tomorrow

The Earth's sea levels are believed to currently be several meters below those reached near the end of the last interglacial cycle (see Figure 1.16 in Chapter 1). If past trends are a pattern, it is certainly possible that sea levels will rise to those historic levels or higher before the end of the current interglacial period. Although such levels have occurred before, such occurrence would be of significant concern to humans since it is without precedent during modern human society. Even though it may be years in the future, if that scenario does indeed happen, there are low-lying coastal areas of the Earth, including a number of densely populated urban areas, that may be subject to flooding. Along both the Gulf and Atlantic coastlines of the United States, evidence suggests that a significant portion of recent reported sea level changes may be related to vertical land motion.

Regardless of the cause, these results together with the possibility of higher rates of sea level change in the future suggest that, in addition to the use of satellite observations at permanent tide stations, coastal planning for sea level change is warranted. Public policies for increased restriction of development activities in coastal zones, although generally not popular public policy, may need to be considered. For consideration in such development restrictions, there is very obvious evidence of highest water level locations during past interglacial periods in some areas. One such example is the Cody Escarpment running along the Northern coastline of the Gulf of Mexico (see Figure 1.15 in Chapter 1). That feature, even though roughly 15 mi or more upland of the current Gulf shoreline, graphically reveals the location of water stands from a past interglacial period. It seems only reasonable that policies that restrict long-term development below such indicators should be considered. Additional limitation of other human activities, such as large-scale groundwater pumping, even in locations far from the sea, should also be considered since those processes have been shown to affect sea level. This is due to the fact that much of such extracted groundwater eventually ends up in the sea, either due to runoff or to the evaporation and transpiration processes (Seo et al. 2023). A number of researchers have attributed sea level rise to global warming caused by human air pollution and other current human practices. That certainly may be a factor and public policy should consider regulation of such practices.

Despite preventative measures, the natural forces that have caused the repeated glacial cycles over the centuries may prevail. Although it may be hundreds of years in the future, if such cycles continue, sea levels may rise to levels believed to have existed at the end of the last glacial cycle or higher before the beginning of the next glacial period. Further, with the current changes apparently due to vertical land motion, there may be even greater changes. As a result, relocation of low-lying coastal settlements may be necessary and should be considered as a part of coastal planning.

Restriction and regulation of coastal development and relocation of coastal facilities will no doubt result in, not only higher values for coastal land but also more emphasis on coastal boundaries. This is the reason for the emphasis on procedures for establishing such boundaries in this writing. Such surveying processes should become an essential part of geomatics education, especially in coastal areas. The linkage between dynamic sea levels and the legal and technical aspects of the constantly moving limits of the sea, recognized centuries ago, still prevails.

References

Cole, G., Hale, J., Ward, D. and Joanas, Z. (2017). Use of Bathymetric LiDAR for Paleo landscape description. *Lost and Future Worlds Royal Society Conference*, Buckinghamshire, UK.

Day, J., Gunn, J., Folan, W. et al. (2023). *Emergence of complex societies after sea level stabilized. American Geophysical Union* 88 (15): 169–170. New York: Springer-Verlag.

Emery, K. and Aubrey, D. (1991). *Sea Levels, Land Levels, and Tide Gauges.* New York: Springer-Verlag.

Ohenhen, L., Shirzael, M., Chandrakanta, O., and Kirwan, M. (2023). *Hidden vulnerability of U.S. Atlantic coast to sea level rise due to vertical land motion. Nature Communications* 14: 2038.

Seo, K., Ryu, D., Eom, J. et al. (2023). Drift of Earth's pole confirms groundwater depletion as a significant contributor to global sea level rise 1993–2010. *Geophysical Research Letters* 50 (12): e2023GL103509.

Index

Sea Levels and Coastal Boundaries, First Edition. George M. Cole.
© 2024 John Wiley & Sons, Inc. Published 2024 by John Wiley & Sons, Inc.